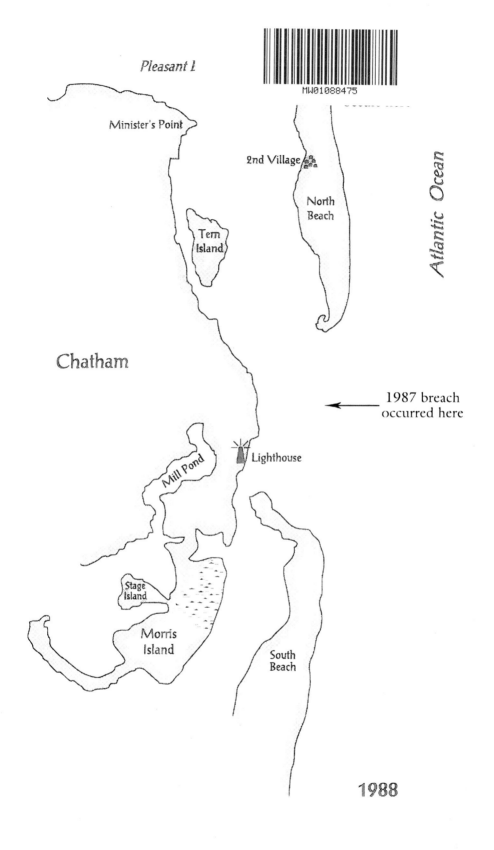

Also by William Sargent

Fukushima, 2012

The Well From Hell, 2011

Sea Level Rising, 2009

Just Seconds from the Ocean, 2008

Crab Wars, 2006

The House on Ipswich Marsh, 2005

Praise for Other Books:

"With his fine descriptions and lucid explanations, Sargent joins the company of Lewis Thomas and Stephen Jay Gould as a first-rate interpreter of modern science."
~ *Publisher's Weekly*

"It is a gem of Natural History… the best introduction to the original environment of the New England coast."
~ Dr. E.O. Wilson, Harvard University

"A joy to read."
~ *The Washington Post*

"A Great Read! Sargent takes us on a raucous jaunt through the New England forest, to see the big picture with unclouded eyes. A true biologist, he examines everything in sight and counts it relevant, connecting it with seamless prose into the rational new picture. It's a powerful boost to the new Nature religion that references us to Life on Earth."
~ Dr. Bernd Heinrich

It's science writing that reads like a novel, with all the page-turning excitement of a thriller."
~ William Martin

"Sargent can turn an event as mundane as a rising tide into poetry. This is a book for everyone who loves the shore, especially Cape Cod."
~ *The Boston Globe*

"If you only have time for one book about life-death dramas played to the sound of crashing waves, about new science and the old sea, about Nobel prizes, squid brains and sex orgies on Cape Cod beaches, then this book is for you."
~ Dr. A.A. Moscona, Journal of the American Medical Association

Beach Wars

Ten Thousand Years of Conflict and Change
on a Barrier Beach

Acknowledgements

In 1938 Bob Thielen, a writer for the *New Yorker*, was vacationing in his summer camp on Martha's Vineyard with his wife, Virginia, and their Jamaican maid, Lucy. By lunchtime the wind was rising and the sky had a curious yellow hue. While Lucy lay out the plates she kept warning about "hurricane weather." But Virginia reassured her, "Hurricanes seldom get this far north."

An hour later the winds were making a continuous booming noise that Virginia thought sounded "like giant kettle drums playing wild crescendos." Fearing that the camp would be swept out to sea, the three quickly donned foul-weather gear and raced outside. They were running along the beach toward a sand dune on the edge of Stonewall Pond when the spray-flecked waters of a storm surge wave swirled over the barrier beach, engulfing their camp. Knowing that Lucy couldn't swim, Bob held her hand as they struggled through the knee-deep waters against hundred-mile-an-hour winds.

Nearby, the artist Thomas Hart Benton was returning from plucking mussels off the rocks as he did every morning to feed his destitute family. As he and his son T.P. approached Stonewall Pond they could see a twenty-foot-high wave of blue-gray water, topped with another ten feet of white spray and spume, hurtling toward the offshore bar. The wave towered over the tiny figures of the Thielens for one brief moment before crashing down upon them. For another moment they could see the Thielens in neck-deep water, before it swept over their heads into Stonewall Pond.

Bob was still holding Lucy's hand, but his waterlogged pants and boots were about to pull them both under, so he took a deep breath and let go of her hand just long enough to dive down and strip off his offending clothes. But when he returned to the surface Lucy was nowhere to be found. He dove several more times but to no avail.

Equally worried about Virginia, Bob swam to her side, and together they struggled against the strong currents and used the thorn-covered tangle of wild roses and bayberries to pull themselves out of the pond and onto the shore. Looking back through the mist the exhausted couple saw their camp floating lopsided in the pond.

The salt spray stung their eyes as they stumbled on through a field and past some cows lying huddled together against the storm. That was when they heard Benton's voice over the roar, "You two—you're damn lucky. Come up to our house. Rita has the fire going. We thought you were lost for sure."

Benton and T.P. helped the shell-shocked couple climb the few stairs up into their house, where Rita met them with open arms and tears pouring down her face. She gave them rum and dry clothes, and sat them down in front of the roaring fire. For a while, all they could do was just stare into the flames, quietly eating Rita's casserole. Bob was still shaking so hard he could barely hold his glass. "Lucy just went down. She just went down. Nowhere to be seen. I should have gone back." Tom and Rita assured him that it would have only been futile and very dangerous.

Eventually, the hot rum and warm fire did their work and Bob started to relax, "I never did like cows up until today." Tom looked at his friend in utter astonishment.

"Now, why in hell do you like cows today?"

"I don't know. It's hard to explain. Seeing them there on the hillside looking so warm and safe," he said quietly, before staring back into the flames.

The scene had been indelibly etched into everyone's brain. Benton made a sketch and a quick study on the Vineyard, then painted a final version of "The Flight of the Thielens" in his studio in New York. "The Flight," along with Bob Thielen's article about the 1938 hurricane, gained Benton national attention. He eventually became the foremost practitioner of what became known as regionalism, an unfortunate title for his new school of American realism. As a WPA-supported painter he went on to trek across the county, sketchbook in hand, chronicling rural life in America just as it was starting to disappear during World War II.

As soon as I realized that there were two versions of "The Flight of the Thielens," I knew I had to have the first version for the cover of this book. I felt the study captured the scene with far more force and immediacy than the more stylized final painting.

My quest to locate the study became a saga in its own right. It started on Martha's Vineyard, where Megan Otten Sargent, the owner of the Gay Head Gallery, first introduced me to Polly Burroughs. Polly had written a book about how important Benton's summers on Martha's Vineyard were in developing his own unique style, so contrary to those of the European-influenced schools of art, centered in New York at the time.

Stephen Sifton of the Thomas Hart Benton Association in Missouri put me in touch with Michael Owen of the Owen Gallery in New York, who finally led me to Henry Schwob in Atlanta, Georgia. Henry and his wife have one of the most extensive collections of American art in the country and they felt exactly the same way I did about the study. Henry had purchased it after his daughter spotted the study on consignment at the Hollander Gallery in St. Louis.

I would like to thank all the people who helped me track down the provenance of the painting, and particularly Henry Schwob and his wife, who gave me permission to use their cherished study of "The Flight of the Thielens" in this book.

Shortly after the 2007 inlet opened in Chatham, Tim Wood, the editor of the *Cape Cod Chronicle* set up an online forum about North Beach. The site became a valuable service to many camp owners and people who just wanted to know about how beaches move and change. It also gave me insights into how intensely people felt about living on the beach and it convinced me that a good book could be written about their battles to save the camps threatened by sea level rise.

As I delved further and further back in time, I realized that their battles were just part of the ever-evolving history of this barrier beach. I would particularly like to thank Tim Wood, Bill and Dan Ryan, Richard Walton, Steve Batty, C. Whiting Rice, Thadd Eldredge, Dana Eldredge, Bill Hammatt, Bob Long, Scott Morriss, Pat O'Connell, Mac MacAusland, Robert Ryder, Rob Crowell and the many others who posted their opinions in the *Chronicle* and on the forum. The New England Biolabs Foundation, the Quebec-Labrador Foundation and the Sounds Conservancy provided much needed funding to do some of my initial research.

Steve Leatherman and Paul Godfrey told me entertaining stories about their early days studying beach dynamics, and Ian Nisbet, Brian Harrington and Scott Hecker filled me in on the Piping Plover dilemma, while Mark Adams filled me in on modern research on the beach. Most of the other people who helped me are mentioned in the book. I apologize if I have left anyone out.

Jill Buchanan did her usual masterful job editing my manuscript and designing the book. It is such a pleasure to have an editor who is so familiar with the nuances of writing about science for a popular audience. Most of all I, would like to thank Kristina and Chappell for putting up with my long hours in the attic spent pounding out these pages, and Ben who has launched his own unique career chronicling the culture and cooking of the many coastal communities that ring our continent.

Table of Contents

Introduction		1
Chapter One	*The Beach, June 13, 2011*	5
Chapter Two	*The Matriarch, Cape Cod 10,000 B.C.*	10
Chapter Three	*The Match, Nauset Beach November 22, 1600*	13
Chapter Four	*Poutrincourt's Oysters Mallebarre, 1607*	18
Chapter Five	*The Sparrowhawk, Nauset Beach June, 1626*	22
Chapter Six	*Never Trust a Bipolar Pirate Eastham, April 26, 1717*	28
Chapter Seven	*Haying the Outer Beach Marsh August 10, 1720*	34
Chapter Eight	*The 1846 Inlet The Wreck of the Orcutt, December 1896*	40
Chapter Nine	*Unraveling an East Coast Secret April 19, 1863*	47
Chapter Ten	*"No Book, No Marriage" Eastham, June 1925*	52
Chapter Eleven	*Rum Runners and Mooncussers December 28, 1929 One Night In July*	57
Chapter Twelve	*Monomoy Island, August 7, 1958*	64
Chapter Thirteen	*The Debate, February 1973*	68
Chapter Fourteen	*The Nude Beach Battle, Truro, August 25, 1974*	74
Chapter Fifteen	*Field Research, Truro, June 20, 1977*	78
Chapter Sixteen	*Coast Guard Beach, February 5-8, 1978 The Blizzard of 1978*	83

Chapter Seventeen	*Piping Plover, May 15, 1986*	**88**
Chapter Eighteen	*The "Old" Inlet, 1987* *The Galanti Cottage January 21, 1988*	**93**
Chapter Nineteen	*"The Perfect Storm," October 30, 1991*	**100**
Chapter Twenty	*Life's a Beach, North Beach, 2002*	**106**
Chapter Twenty One	*The New Inlet, April 24, 2007* *The Attack, August 11, 2007*	**110**
Chapter Twenty Two	*Hurricane Noel, November 3, 2007*	**117**
Chapter Twenty Three	*The Last Summer, North Village, 2009* *The Hammattyville Gale* *Hubris, October 18, 2009*	**124**
Chapter Twenty Four	*How to Predict Erosion*	**129**
Chapter Twenty Five	*The Beginning of the End, North Beach Island, August 1, 2011* *Legalities*	**134**
Chapter Twenty Six	*Winning the Battle but Losing the War March, 2012*	**139**

The Barrier Beach.

Introduction

2012 marks the three hundredth anniversary of the town of Chatham, where most of the events in this book take place. Throughout this history, and for thousands of years before, Chatham's inhabitants and her barrier beach have been dealing with the effects of sea level rise. In this book I have tried to put Chatham's recent sea-level-rise problems within the context of the ten thousand year history of the barrier beach that protects her mainland.

Throughout the 1600's and 1700's people only ventured out to the barrier beach to hunt, fish, mooncuss and lop off the bejeweled finger any pirates who happened to run aground, so the locals hardly noticed the effects of the rising seas.

But during the 1800's people started building houses on the

1

mainland protected by the barrier beach. When an inlet opened up in 1846, Chathamites certainly noticed when their lighthouse started pitchpoling down the front of its coastal bank and the streets and homes of the village of Scrabbletown were washed into the foaming white froth of the Atlantic Ocean.

By the 1900's people on North Beach had established villages of permanent camps that they reached by driving down the beach from Orleans. The outer beach quickly became a place where families could gather and local kids could blow off steam by crashing twenty-five-dollar jalopies into each other without disturbing the more sedate inhabitants of the mainland. By the time the inlet opened again in the 1980's people felt they had the right to drink, drive and live in permanent houses on the barrier beach.

It was a sad and wrenching experience for Chatham when a "new" inlet broke through the barrier beach in 1987. The town was engulfed in chaos and lawsuits as people tried to build seawalls and revetments to protect their mainland properties.

Town and state officials learned that any regulation designed to deal with sea level rise is bound to have flaws and if they were applied too strictly the implementation of those laws could look arbitrary and capricious.

In an attempt to rectify those problems Chatham hired an "erosion czar," a single person to manage all the town's marine resources. It started spending a large part of its budget dealing with the effects of sea level rise. Town and state officials also decided to lean over backwards to accommodate people facing the future loss of their homes.

The results were no better. When a second inlet burst through the barrier beach in 2007, camp owners spent close to half a million dollars moving their homes multiple times before they too were swept away by the whims of Mother Nature.

Then, in 2012, when the Cape Cod National Seashore felt it had to act proactively to remove five camps owned on North Beach Island, it almost got it right. But it also learned that you can't just rely on long term rates of erosion to make fair and sound decisions about removing someone's beloved summer home.

At the same time all these difficult events were taking place, the new inlets were also making the waters behind the barrier beach cleaner and more productive. The employment rate in Chatham actually increased by almost one percent as hundreds of shell fishermen discovered they could make a good living digging clams on the newly-productive clamflats.

Most of all, the entire town and its hundreds of thousands of summer visitors were able to witness the incredible power of nature to shape and mould our planet, and they were able to do this on one of the most beautiful spots in the world.

These are lessons the three million seven hundred thousand U.S. citizens who live less than a meter above the sea, and the two hundred other barrier beach communities will be wise to heed in the coming decades.

With these thoughts in mind this book is dedicated to the people of Chatham, who have learned so much from living on and beside this ephemeral and ever-changing barrier beach.

Musells increased dramatically after the 1987 inlet opened.

The Great Beach. Photo by Ethan Daniels.

Chapter One
The Beach
June 13, 2011

The Great Beach of Cape Cod shimmered in the light of the full moon, a stretch of sand as much ocean as land. The quiet huffing of whales could be heard feeding just beyond the surf line. Tomorrow, a great white shark could attack an eight-hundred-pound gray seal twenty feet from this shore. Only half a mile from the bright lights of the mainland, this sliver of sand and marsh was as close to the wilds of nature as the peaks of the Himalayas or the savannahs of East Africa.

Similar beaches stretch along the East Coast, from New England to Florida and along the Gulf of Mexico to Padre Island, Texas. They make up the longest barrier beach system on our planet and are natural breakwaters, crucial to protecting over a million people from the ravages of sea level rise.

These beaches have been growing for millions of years. Their sand comes from the gradual breakdown of the craggy mountains of New England, the gentle foothills of the Mid-Atlantic Piedmont and the billions of mollusks, coral and curious-shelled bacteria that thrive off our southern coasts.

When water freezes and thaws in cracks of our continent's bedrock, it releases tiny, almost indestructible grains of minerals:

, mica, and even minute gems of semi-precious by year these minerals tumble down the sides of ct in rivulets and are swept down roiling rivers ck and Connecticut in New England, the East 's in New York, the Delaware and Susquehanna ... the mid-Atlantic, and the mighty Mississippi on the Gulf of Mexico.

Billions upon billions of mollusks also thrive in the warm shallow waters off Florida and on the flats of the Bahama Banks. Diving underwater, you can feel the vibrations of nurse sharks as they use their stomach muscles to crush shells, and you can hear the fused-toothed beaks of parrotfish as they tear great mouthfuls of calcium carbonate off the coral reefs. The parrotfish digest the coral animals then excrete the grainy detritus in great clouds of pure-white, ready-made, calcareous sand. These grains are then pushed and rolled around by waves until they finally wind up on beaches during storms and high tides. There, they will be worked over some more by the winds and weather, as they have been throughout the eons.

About a hundred fifteen thousand years ago the ellipticity of the earth's orbit, the precision of the equinoxes and the gravitational perturbances of nearby planets, together called the Milankovitch cycles, worked in synchrony to initiate the last great Ice Age. Thirty thousand years ago the Ice Age was at its peak and the world's oceans were about three hundred feet lower than today. The glaciers pushed as far south as the islands of Martha's Vineyard, Nantucket and Long Island. Florida was about twice as broad as it is today, and the beaches and coasts of the mid-Atlantic were several miles east of their present location.

Eighteen thousand years ago the world was rapidly emerging from the Ice Age. The seas were rising almost three feet every fifty years, or about six times faster than today's pace. In places, the seas were rising so fast that sixty feet of land would be inundated in a single year. Florida sank quickly to half its former size, and most of the events that left places like Cape Cod in their present form happened in less than a thousand years.

Milutin Milankovitch was a brilliant Serbian astronomer who calculated in the 1940's that these orbital cycles caused climate change. His theories were only proved correct in the Seventies when oceanographers discovered that fossil samples of warm- and cold-water species of plankton corresponded precisely with the orbital cycles he had discovered thirty years before.

The glaciers left behind the moraines, drumlins, beaches and eskers of New England. Perhaps the most famous moraines are the curving ridges of sand, gravel and rocks that make up the uplands of Martha's Vineyard, Nantucket and Cape Cod. The drumlins are low hills of glacial till that were left by the glaciers as they retreated north. Eskers are low ridges of sand that snake across the New England landscape; they were deposited by the many interlacing rivers that drained the glaciers during their rapid retreat.

Waves continued to work on these rugged piles of glacial till. As they attacked the cliffs and promontories they winnowed out sand that would collect into long, growing barrier beaches parallel to the mainland.

About five thousand years ago the pace of sea level rise had slowed to a consistent rate of about a foot every hundred years. By then, the East Coast and Gulf Coasts had their present configurations of productive bays, fertile marshes and low coastal prairies that are refuges for wildlife, and are protected from severe storm damage by a series of over two hundred barrier beaches.

Nauset Beach is one of those barrier beach systems. Its beaches run from Eastham to the southern tip of Monomoy Island in Chatham, Massachusetts. It is the largest barrier beach system on Cape Cod and one of the best-studied systems in the world.

Under the influence of the slowly rising seas, the Nauset Beach system has been gradually migrating landward for the past ten thousand years. Throughout this time it has protected the shallow waters of Pleasant Bay and provided sustenance to the plants, animals and humans of this fertile region.

In recent years we have learned that New England's prevailing northeast winds push water up against this beach, creating "longshore currents" that travel parallel to the shore. On calm days small waves pull sand off the beach in these longshore currents until the next wave washes them back ashore a few inches down the beach in a looping pattern. Almost imperceptibly, these currents wash more than a dozen railroad cars worth of sand down the beach annually, causing it to grow more than half a mile longer each year. It is one of the most dynamic geological features of our planet.

Storms can change the beach much more rapidly. In a process called "rollover," a strong winter storm can carry as much as sixty feet of sand from the front of the beach over the dunes and into the backshore marsh. After the storm passes the beach rebuilds, but it will have migrated several feet closer to the mainland.

If you were to take a time-lapse photo of a barrier beach every day, you would see that it constantly shifts and pulsates like a living

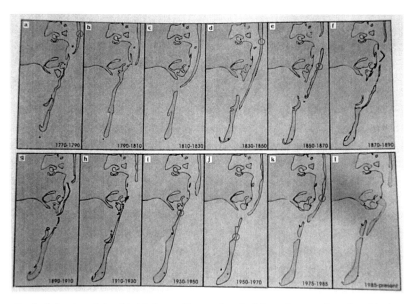

Cyclic behavior of the tidal inlet at Nauset Beach, Massachusetts. Developed by by Elizabeth A. Pendleton, after Graham S. Giese, 1988.

organism, but it seldom erodes totally away. By maintaining its form, it can continue to act as a living barrier that protects the coasts from offshore storms.

In recent years researchers have also discovered that, in Chatham, a new inlet forms about every hundred forty years, creating a three-to-six-mile-long barrier beach island. Because this island is cut off from its upstream source of sand it rolls over at the rate of almost a hundred feet a year, dooming structures in that path.

Sea level rise is the force driving all these processes. The seas have been rising for the past ten thousand years, but climate change has increased the pace of sea level rise in the past twenty years, and will continue to do so during the coming decades.

This process has been going on for over ten thousand years and for over about nine thousand five hundred of those years humans have been content to simply visit our planet's barrier beaches for food and respite. It has only been during the past fifty years that humans have had the temerity to think that they could actually build permanent structures and even cities on such fragile stretches of ever shifting sand. This has made recent controversies on the beach more intense.

In this book we will focus in on Nauset Beach to show how such a natural system has been affecting the lives of people for these thousands of years. It is one of the best places in the world to observe how such rapidly changing geological features can effect our lives and the ecosystems we depend on.

We start when the glaciers were melting rapidly and a dominant group of mammals received their first taste of what it would be like living beside a curious new species of bipedals that called themselves humans…

Mastodon. Kristina Lindborg.

Chapter Two
The Matriarch
Cape Cod 10,000 B.C.

It was twelve thousand years ago. A herd of mastodon was roaming slowly across what today we call Cape Cod. It was hot and the mastodon were plagued by swarms of flies and encumbered by fur evolved for a far colder climate. The coast was three miles east of where it would be in the year 2012.

The matriarch of the herd had seen many changes. Every summer the thin strand of sand between the ocean and icy cliffs grew. The Wisconsin glacier, which had once stretched all the way to what we now call Martha's Vineyard, had retreated north to what we now know as Maine, and it was continuing to melt northward. The matriarch had seen the sea level rise nine inches in her lifetime. Every year she saw sixty more feet of coast succumb to the rising sea. Change was everywhere.

The sea had been rising so fast that nature had not had time to build up the barrier beaches, salt marshes and lagoons that are hallmarks of a gradually-rising sea. Without these features, biological diversity along the coast was low. There were few of the birds, fish and mollusks that depend on the salt marsh food chain.

The roving bands of Paleo-Indians who had recently migrated into this area still mostly hunted in the inland areas. So, as long as the mastodon herd stayed on the narrow shore, it would be safe from the humans who might find them a hearty meal.

But the matriarch was faced with a dilemma. She knew it was safer to stay on the beach, but the herd was hot and deeply annoyed by the greenheads and mosquitoes.

Even her own calf was running around her legs, flapping his ears and trumpeting his annoyance. The matriarch was none too comfortable herself. The insects seemed to be worse than anything she could remember. Finally, she decided. She lifted her trunk and smelled the air for signs of water. She remembered a glacial lake that her mother had taken her to when she had been the herd's matriarch. The cooling waters had relieved the herd then; perhaps they would do the same again.

With a determined air, the matriarch wheeled and the herd followed. Excitement rippled through the older animals. They trusted she would save them from their miseries. With much jostling and trumpeting the herd trotted up the bluff and onto the open upland.

A few hours later the herd spotted the glacial lake shimmering through a conifer forest. The mastodons broke into a gallop and splashed straight into its cooling waters. Soon they were happily gamboling about and spraying each other with their long prehensile trunks. The matriarch's calf cavorted rambunctiously around his mother before spotting a log drifting on the lake's quiet surface. While his mother watched the herd, the calf swam over to investigate.

Suddenly, a great shout erupted from the shore. Six men rose from the log and splashed furiously toward the infant. Other Paleo-Indians rose out of the sedges and launched more logs from the far shore. On land the mastodon can more than hold their own, but in the water they are slow and vulnerable.

The matriarch tried to regroup the herd, but lost sight of her calf. Two logs had cut him off from the others, and now the men were closing in from both sides. If the infant kept swimming he may have escaped his pursuers. But instead he panicked and turned back toward his mother.

The first spear glanced off his back, but the second drew blood. Now the calf was trumpeting in pain and mounting terror. The matriarch heard his pleas but could do nothing. Another spear found its mark, and blood spewed into the roiling waters. One man started clubbing the infant with a heavy stone axe. Each hit drew more blood.

A shudder passed through the calf's body and he swam in jerky circles. One of the warriors jumped onto to the calf's back and started to hack great hunks of flesh off his exposed flanks. But the added weight of the man was too much. The infant's body slipped below the surface and out of sight. On shore, the matriarch bellowed her rage and sorrow. It was the first time she had encountered humans. She would not lead the herd back to this lake again.

The infant's body drifted slowly toward the bottom, where it would decompose and be covered with sand and mud. Thousands of years later the infant's partially-mummified body would erode out of the cliff and tumble down onto the beach once again. But it would be a new beach, three miles shoreward of its former location.

Nauset Village. Kristina Lindborg.

Chapter Three
The Match, Nauset Beach
November 22, 1600

The autumn sun rose slowly out of the Atlantic Ocean to shine on the tidy fields and open forests of the Nauset village. The tribe was at the peak of its influence. Food was plenty, trade flourished and the sachem had kept the nation out of war for as long as anyone could remember.

A young warrior stepped out of his long house to greet the newborn day. He was a tall, quiet man with clear, bronzed skin. At his feet lay a pile of oyster shells, the remains of last night's feast. Attaquin swept the shells into a reed basket, walked to the marsh edge and dumped them there before reclining on a pile of deerskin blankets. He enjoyed sitting in the sun, watching the village before it was fully awake.

He looked out over the fields covered with the stubble of this summer's corn. The harvest had been good.

Beyond the fields were open woods. Every year the warriors burned the underbrush to keep the forest open so they could hunt deer and turkey in the tamed landscape.

Below the village lay the quiet water of Portanamaquut Bay. This biologically rich bay provided the Nausets with the quahog, conch, striped bass and bluefish that made up the bulk of their nutritious diet.

Scores of other villages hugged these shores that would soon be called Cape Cod and Pleasant Bay. Now it was simply a loose confederation of villages within the Wampanoag nation.

As Attaquin walked back from the salt marsh, a runner from the Monomoyicks was just arriving.

"Attaquin, can you join us? Some whales have come ashore on the great beach. Is your dugout ready?"

This was the moment Attaquin had been waiting for. For many months he had carefully hammered and chipped his knives. Now they were thirteen inches long with beautiful fluted edges. They looked like fine ceremonial objects, but Attaquin had other plans for his finely-wrought tools.

Attaquin and his brother were known as the best swordfishermen in the village. They had spent many moons building a dugout canoe for fishing. Starting with a thick pine tree they had found on an inland journey, they laboriously hacked away at the interior of the log with adzes and smoked it with small fires. When finally done, their dugout was long, wide and sturdy enough so the brothers could haul it over the beach and paddle far out into the Atlantic where the great fish slumbered on the surface.

Attaquin and his brother had learned to work as a deadly efficient team. Attaquin would stand in the bow, giving hand signals so Quitsa could quietly paddle up behind the swordfish, unsuspected. The men would hold their breath as Attaquin quietly slipped his spear into an atlatl to help it fly further. He would thrust the spear deeply into the muscular back of a somnolent giant. Then, they'd throw out balloons made from the bladders of deer to slow the fish down until it eventually tired enough so that Attaquin could deliver the coup de grace.

The entire village would assemble when the brothers returned with their catch. Attaquin would clean the fish, and Quitsa would distribute pieces to every member of the band. He did this skillfully, garnering praise and political obligations. The sachem Paysum would thank the brothers, the elders would praise them, and all the young women would vie to catch their attention.

Now this rare discovery of whales would give Attaquin and Quitsa another chance to provision the village and display their prowess. With rising excitement Attaquin wrapped his long knives in deerskin, packed them carefully in the bow and pushed the dugout into the bay.

Soon dugouts from other villages joined the brothers on the bay. The Nausets paddled together down a creek beside Paysum's Island and hauled their dugout over the beach to the broad waters of the Atlantic Ocean. A mile down the beach they could see the dark forms of the whales stranded on the sand flats. Each carcass was surrounded by a half dozen men hacking at the whale's tough skin. As they approached Attaquin called out.

"I am Attaquin of the Nausets. We have come for our share of the whales."

"What took you so long, Attaquin? The whales are almost gone. Yours are the scrawny ones down at the end of the beach." It was Pashto, Attaquin's old friend from the Monomoyick tribe.

"How are you Pashto? Don't tell me you are in charge of the whales, we'll never get any. Is your sister here to see how real warriors butcher a whale?"

"Nananatuck is waiting. She plans to dance with you after the whales are butchered. But by that time you will have lost your loincloth to some real warriors. Did you bring your sticks? Tonight we start the first ball game; Nausets against the Monomoyicks, winners get the girls, losers lose their loincloths!"

With such bantering Attaquin and the Nausets paddled down to the section of beach reserved to them by treaty. Soon they were busily stripping great slabs of muscle-rich red meat off the whale's bloody carcasses. Warriors from other villages wandered over to admire Attaquin's knives that cut so deeply through the thick layers of blubber and meat. Attaquin could dress out two whales to their one.

"How much do you want for one of your long knives, Attaquin?"

"More than you can afford, Pashto!"

"Slow Duck here has a knife he traded with a white fisherman who sailed to our village on the last full moon. We don't like those white devils, but they keep coming back."

"You were probably too polite to them."

"Yes, we made a big mistake. We showed them what good food we have. Now they won't leave us alone. Have you ever eaten the food the Englishmen eat? I wouldn't feed it to a wolf. We ate some of the white man's food to be polite, now look at us. We are tired and weak, and Slow Duck has just come down with red bumps all over his body.

After the butchering was over, the villages separated into two teams. The warriors laughed and jostled each other as they battled with sticks and their bodies to drive a small ball up and down several miles of beach. The games continued for four days. Afterwards, the warriors hung clothes, spears and wampum on a driftwood arbor and rolled pieces of deer antler to gamble on each other's wares. The point of the contest was to bet, laugh and swear loudly at the outcome. On the last night the villages danced and Attaquin slept with Pashto's playful sister Nananatuck.

The following morning Attaquin and the rest of the vanquished Nausets climbed into their dugouts to paddle back to their village. Nananatuck and her friends lined up on the top of the dunes to laugh at their new friends as they paddled away, naked.

Still, the Nausets congratulated themselves. It had been a good clean beach battle, but perhaps too clean and not quite good enough. They vowed that next year it would be the Monomoyicks who would lose their loincloths.

After several hours of heavy paddling Attaquin and his brother returned to their village. They distributed the whale meat to the sachem and waiting villagers, and then repacked the dugout with spears and Attaquin's new flencing knives. They covered the dugout in deerskins and buried it under earth and cedar branches. It would be waiting for them, packed and ready to go, when the whales, striped bass and swordfish returned in the spring.

Little did the Nausets know that the small pox Slow Duck had picked up from the white man's knife would race through their village and nobody would be alive to unearth the dugout the following spring. It would remain buried for the next three hundred fifty years.

Poutrincourt's Oysters.

Chapter Four
Poutrincourt's Oysters
Mallebarre, 1607

Samuel de Champlain and Jean de Poutrincourt were in good spirits on September 30, 1607. The explorer and the Governor of the Arcadia area were exploring new lands to establish another French settlement. Suddenly, one of the harquebusiers shouted they had discovered some oysters inside Cape Cod Bay. The Frenchmen deemed them almost as good as any found in France. Today we call those oysters "Wellfleets," and continue to pair them with the finest Bordeaux's.

The explorers named the harbor "Port aux Huistres," then headed for the Nauset Indian village they had visited the year before. The Nausets of "Port de Mallebarre" had promised to grow extra corn for the Frenchmen on their return.

High surf was breaking over the bar that protected what we now call Nauset Harbor. Several Indians paddled through the waves and gestured to Champlain that if he sailed all the way around the dangerous beach, he would find a large Monomoyick village on the southern shore. But sailing around Mallebarre was easier said than done, and in their attempt the explorers damaged their rudder in Pollock Rip.

After making safe harbor, Champlain extended his telescope and scanned the shore. "Il as raison," he murmured to Poutrincourt. The bow watchman had seen smoke rising from several long houses arrayed around a sheltered harbor. The fields were covered with the stubble of the recent harvest and the forests had been burned to make it easier to hunt without all that underbrush.

The Indians were having some kind of celebration. There were at least a hundred fifty Monomoyicks singing and dancing on the beach. This made the crew very happy, indeed. They remembered how tasty the clams and lobsters had been at the clambakes they had attended at Patuxet the year before. Everyone was in a good mood as the Indians paddled out to greet them.

Champlain gave orders that Poutrincourt and the harquebusiers be ferried to shore to explore the inland areas; he sent a small contingent of sailors to the beach to barter for food and repair the broken rudder. On his return, Governor Poutrincourt told Champlain of his great delight at finding so many cypress, oaks, pines, and nut trees. He was especially pleased they had found so many fine grape vines, but of course they were certainly not as good as those found in the famed Loire Valley.

That evening, Captain Champlain wrote in his log, "This would prove a very good site for laying and constructing the foundations of a state, if the harbor were a little deeper and the entrance safer than it is."

But the Monomoyick had not taken kindly to the harquebusiers snooping around their land. And what about that strange wooden cross the French had erected in the sand after chanting quietly in unison and bowing and scraping in the sand? What was that all about? For their part the French wondered why the Indian women were taking down the village wigwams. Was it because the men were about to attack? Nobody could convince the Frenchmen that the Monomoyick took down their long houses every year in order to move to their inland hunting grounds.

Just to be on the safe side, Captain Champlain ordered his crew to return to the ship. But three of his men lingered behind, How could they possibly return before the crust on the bread they were baking had turned a buttery golden brown? Champlain dispatched two more sailors to retrieve them, but to no avail. The inducement of fresh bread was too much to resist. Besides, what could possibly happen if they simply enjoyed some freshly baked bread and returned to the ship in the morning?

Four hundred Monomoyick surrounded the recalcitrant bakers at daybreak and "sent them such a volley of arrows that to rise up was death." Four of the sleeping gastronomes were killed outright. The last was allowed to live to tell the tale.

The Frenchmen returned to try to capture some Indians to use as slaves at Fort Royal, their new colony in northern Canada. But the Indians proved to be too swift and wily. Seven Monomoyick were killed, but the French never returned to try to settle on the New England coast again. Champlain named the sheltered harbor, "Port Fortune, on account of the misfortune which happened to us there."

A few year later, English Captain John Smith's lieutenant, Thomas Hunt, dispensed with such Gallic appreciation for nuance and food when he lured twenty Indians aboard his ship, then sailed off to Malaga, Spain, to sell the poor souls into slavery.

One of those Indians was Tisquantum, whom we have come to know as Squanto, the paragon of Native American hospitality. Several of Squanto's owners admired his many abilities, and finally brought him to London where the explorer Thomas Dermer signed him on as interpreter on his trip to New England.

When Squanto finally made his way back to his village in 1619, he discovered that it had been decimated by smallpox. If he had stayed in New England, he also undoubtedly would have succumbed to the epidemic. Instead, he had probably picked up immunity to smallpox by being exposed to a milder strain of the disease during his travels in Europe.

When the Pilgrims arrived in Patuxet a year later, Squanto decided to join them. It was not a difficult choice. The Wampanoag nation had been wiped out by smallpox and Patuxet was empty. Besides, he had no job and the Pilgrims needed a translator for the settlement they would call Plymouth.

But there was a darker side to this patron saint of Thanksgiving. In dealing with his fellow Indians and competitors, Squanto was not above hinting that the Pilgrims kept smallpox buried in the ground and could call it out at will.

The English didn't entirely discredit the idea. When a young native warrior asked about a suspicious bag of powder that he had seen the Pilgrims bury, the Englishmen assured him it was gunpowder, not smallpox, "But our God does have powere over the plague and can use it against our enemies." It was not the first time that someone had claimed God was on their side, and it would not be the last time that an entire village would be wiped off this coast by an invincible force of nature.

The Sparrowhawk on Boston Commons.

Chapter Five
The Sparrowhawk
Nauset Beach, June, 1626

Captain Johnston lay in the dark, groaning with agony. He was about to lose another tooth from the scurvy that had plagued him for the last few weeks.

"Good Morning, Captain." It was the first mate bringing him yet another bowl of cold gruel.

"What is it, Snipes?"

"We are out of water, beer and woode, sir."

"Of course. And what of the passengers?"

"Sipsie's redemptionists are mad for drink and the crew is mad for land."

Captain Johnston groaned again. The symptoms were getting worse. It would be over a hundred years before the Admiralty figured out you had to give fresh fruit to your sailors to prevent them from getting scurvy. There were no "limeys" aboard the Sparrowhawk. The crowded, forty-one-foot barque had been at sea for almost six weeks, and the captain was still not sure which way to go to reach the Virginia colonies.

"Then the die is cast, Snipes. We shall steer between the northwest and southwest, caring not whatsoever land we come upon."

Finally, something more definitive than a few desultory orders from Johnston's darkened cabin, thought Snipes to himself.

"Good to hear it Captain. Glad you are feeling better. The crew will be much delighted."

The crew's delight would not last long. A few days later the bow watchman spotted land. It was not a lightly forested headland of fields and valleys, but the long white stretch of dangerous sand protected by offshore bars and growling surf that the French had named Mallabarre. Snipes knew the Sparrowhawk's small twin topsails would not provide sufficient headway to keep them off the outer bar.

The captain ordered the crew to lay out a single anchor. But its cable snapped when the winds rose during the night, and it looked like the wayward colonists surely would be pounded to pieces by daybreak.

Yet, the following morning found the Sparrowhawk firmly ensconced in the crystal white sands of Monomoyick Bay. The high tide course had lifted the thirty-ton vessel over the bar and floated her gently inside the harbor where she grounded in the sand opposite Portanaumaquut Village, almost a mile away.

Attaquin's great grandson held his hand over his brow and squinted into the rapidly rising sun.

"Nontumsa, come quick. It looks like the sails of one of the white man's boats."

"It can't be."

"Surely, it is. Their cargo must be lying in the sand."

"No, you must go wake sachem Paysum and see if you can help."

Snipes ordered his crew to stand guard as the dugouts approached.

But Attaquinna hailed the astonished colonists in their own language.

"Hello, Englishmen. Ye Governor William Bradford is a good man. He lives across the waters in Plimouth. Our runners can reach him in a few days. Perhaps he can help you."

Mr. Fells and Mr. Sipsie could not believe their good fortune. They wrote a letter to Governor Bradford requesting spikes and oakum to repair the Sparrowhawk and ordered two of their men to accompany the Indians across Cape Cod Bay to New Plimouth.

Governor Bradford returned with the men by shallop to Naumiskakett Marsh. They portaged the supplies over the mainland to Portanaumaquut, and then borrowed a dugout to paddle across Monomoyick Bay to where the Sparrowhawk lay stranded on the inner beach.

Several days later Captain Johnston had the vessel repaired, and he set sail once again for Jamestown. But it was not to be. Another storm drove the Sparrowhhawk back into the sand, forcing the oakum out of her seams and damaging her beyond repair.

This time Governor Bradford took the colonists back to New Plimouth for nine long months. The wayward Virginians proved to be not to the Pilgrims' liking. They drank too much and had far too many domestic infelicities. They were asked to leave the following spring when two ships, bound for Virginia, arrived from England.

Governor Bradford dutifully recorded all these incidents in the Plimouth Plantation log, though he delicately neglected to mention the exact nature of Fells' infelicities, and omitted the details of the many drunken escapades of Mr. Sipsie's Irish redemptionists. Eventually, the entire incident was forgotten and even Cape Codders couldn't remember exactly why they called the stretch of the beach that runs south of Strong Island "Old Ship Harbor."

It took several years, but the Sparrowhawk eventually was covered by blowing sand; marsh grass grew up and over her bulkheads. Finally, the last of her top deck disappeared below the marsh and she lay in an oxygenless crypt of peat and sand—perfect conditions to preserve the wood of her elm- and-oaken timbers. One observer noted that, "successive generations of Doanes would swing scythes and toss marsh hay above her forgotten grave, until these rocky wooded islands shall slip below the trampling surge in pools of sparkling sand."

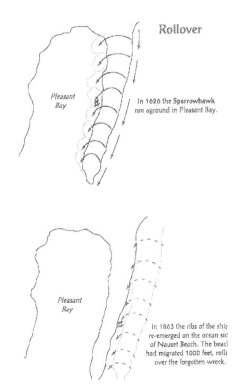

Rollover as described in my book Storm Surge; A Coastal Village Battles the Atlantic.

Chapter Five   25

But the beach would not leave the Sparrowhawk alone. The beach continued to erode and roll over itself. Then, in 1863, a particularly fierce nor'easter tore sixty feet off the front of the beach, leaving the bones of the Sparrowhawk in the surf once again. This time, the sepulchral wreck was on the ocean side of Nauset Beach. How did it get there?

During the two hundred thirty seven years between its initial shipwreck and reexposure, the rising ocean had caused Nauset Beach to completely roll over itself, migrating a quarter mile closer to the mainland. The wreck of the Sparrowhawk had simply stood still while the beach rolled over on top of it, until she had finally come out on the other side.

The discovery of the long-forgotten wreck caused a sensation. Hundreds of Cape Codders flocked to the site to see it. The observers tittered at opium pipes still lying below decks and wondered if they had had a role in the demise of the long-forgotten vessel.

People started turning the timbers into souvenirs until calmer heads prevailed. The wreck was hauled above the high-tide mark, and the famous Boston Naval architect Dennison J. Lawlor and two Orleans shipbuilders by the names of Dolliver and Sleeper were commissioned to transport the wreck to Boston and reassemble her on Boston Commons.

The shipbuilders almost got it right, except that they thought the Sparrowhawk could have only carried a single large triangular sail, as on lanteen-rigged boats in the Indian Ocean. They gave her one mast, set amidships, but she undoubtedly should have had two masts, set fore and aft, to carry her two small twin topsails.

The wreck created yet another sensation in Boston. Spectators marveled at the rake of the barque and how the stern projected so far aft that, although the vessel was forty-one feet along the deck she was only twenty-eight feet on the waterline. The ship was approximately the same size as the Mayflower that had brought one hundred two passengers to Plymouth, and the Arbella that had brought about sixty passengers to Boston. How had the wayward Virginians survive six weeks in such cramped quarters?

Today, the remains of the Sparrowhawk reside in the darkened confines of the Cape Cod Maritime Museum, shuttered in Hyannis. It is an ignominious fate for such an important piece of history, the only existing example of the type a vessel colonists used to sail to the Americas during the Mayflower era. But there she sits.

Who would have guessed that almost four hundred years after her demise, the Sparrowhawk would surface once again to become a bone of contention between those who deny and those who decry global warming?

In search of pirate treasure. Kristina Lindborg.

Chapter Six
Never Trust a Bipolar Pirate
Eastham, April 26, 1717

Captain Sam Bellamy made his way unsteadily toward the bow of the Whydah. He had been celebrating with several casks of Madeira they had captured off Virginia. Tomorrow he hoped to be in bed with Marie Hallett, but first the handsome young pirate had to lead this damn rag-tag fleet of prize ships into Provincetown Harbor.

Sam felt good, too good in fact. His had already been an adventuresome life. He had been born in Devonshire in 1692 and from that day forward had made it his life's ambition to escape the dreary, gray English winters and make his fortune in the colonies. First, he joined the Royal Navy, which had fetched him up on the shores of Cape Cod. There he fell in love with his mistress, the beautiful Marie Hallett of Eastham.

He soon discovered that the long cold Cape Cod winters were no more pleasant than those of England. So he convinced his friend, the Boston goldsmith Paul Williams, to finance a trip to Florida so they could salvage the Spanish Plate Fleet that had gone down in the devastating hurricane of 1715. The storm had drowned a thousand Spanish sailors and left over a hundred million dollars worth of gold and silver scattered on the ocean floor.

The two young men soon discovered that salvaging was sloggingly hard work, but that pirating was fast and profitable. They purchased two long, pirogue-style dugout canoes and set themselves up as freelance pirates with a pick-up crew of disaffected British sailors and Indians from the Miskito tribe of Central America.

The fearless young adventurers amused the veteran pirate Henry Jennings, who had been commissioned by the Governor of Jamaica to plunder the Spanish as they tried to salvage the treasure from their doomed fleet.

Jennings asked Bellamy and Williams if they cared to assist him in the capture of the Sainte-Marie, a French frigate lying off Cuba. After a long and gruesome battle, Bellamy's crew walked away with their share of the Sainte-Marie's thirty thousand pieces-of-eight.

This was more to their liking. Their exploits caught the eye of another pirate, Captain Paul Hornigold of the Mary Anne, who also invited the fearless young man and his crew to join his ship. For Hornigold, it was not a good move.

Bellamy soon became impatient with Hornigold's refusal to capture English ships. He used his silver tongue to convince the crew of the Mary Anne to vote Captain Hornigold off the ship and make him their captain. Now, the two young pirates finally had a ship of their own.

But Williams soon started to have misgivings about his friend. Sam was displaying some of the signs of the climatically-induced hypermania that would continue to plague his otherwise-brilliant career. He now wore long, deep-cuffed cloaks with silken stockings. He carried four large guns and had a sword strapped to his belt. Unlike the veteran pirates, he kept his raven locks in a long black ponytail instead of the foolish looking wigs he had so hated in the Royal Navy. The crew loved their colorful captain and also started sporting gem-studded rings and jewelry, as if to mock the stuffiness of the British colonialists. They were the rock stars of their era and vowed to follow "Black" Sam Bellamy to the ends of the earth.

The crew's next exploit was the successful capture of the Sultana. Afterwards, Bellamy made the Sultana his flagship and gave command of the Anne Marie to Paul Williams. Bellamy was now branding himself as the Prince of Pirates, a sort of philosopher king who was not just out to capture gold, but wanted to build up his own fleet of ships and even establish a haven where pirates could live according to their own democratic rules, outside the sovereignty of any established nation.

Bellamy's greatest success was the capture of the slave ship Whydah off the Bahamas. It took three days and all of his cunning, but the Sultana and the Anne Marie finally overtook the Whydah and raked her with several shots of iron, nails and cannon balls before boarding and capturing her captain, Lawrence Prince.

Because Bellamy now considered himself a gentleman, he graciously gave Captain Prince the Sultana and allowed the vanquished captain to sail away with most of his former crew. He also offered the few slaves left on the ship their freedom if they would join him as pirates; thirty of the men agreed.

Bellamy kept the fast and multi-gunned Whydah for his own flagship as he continued to capture a total of fifty ships laden with treasure before heading back up the East Coast for his rendezvous with Marie. One hundred and eighty bags of gold, silver and jewelry were packed neatly in sea chests below the Whydah's decks. Each crew member would get his own fair share when they arrived in port. Sam could imagine the sparkle of excitement in Marie's eyes when he hung a giant Columbian emerald around her swanlike neck.

In April he captured a wine sloop off Block Island, captained by Simon Beer. When his crew decided they had to sink Beer's ship, Bellamy expressed his sympathy.

"I am sorry they won't let you have your sloop again, for I scorn to do any one mischief when it is not to my advantage; damn the sloop, we must sink her, and she might be of use to you. Though you are a sneaking puppy, and so are all those who will submit to be governed by laws which rich men have made for their own security…

They vilify us, the scoundrels do, when there is only this difference; they rob the poor under the cover of law, forsooth, and we plunder the rich under the protection of our own courage. Had you not better make, then, one of us, than sneak after these villains for employment?"

"My conscience will not allow me to break the laws of God and man," replied Beer.

"You are a devilish conscience rascal, I am a free prince, and I have as much authority to make war on the whole world as he who has a hundred sail of ships at sea and an army of a hundred thousand men in the field; and this my conscience me: aye but there is no arguing with such sniveling puppies, who allow superiors to kick them about the deck for pleasure."

The crew heartily applauded Bellamy's infamous "Free Prince" speech. But Williams' misgivings about his friend only grew more intense. When they were off the coast of Cape Cod he asked if he could take the Anne Marie to visit his family in Rhode Island while Bellamy was visiting Marie Hallett.

As they approached eight bells—midnight in land lubbers terms—Sam sensed that something wasn't quite right aboard the Whydah. He offered Captain Beer his freedom if he would pilot the Whydah safely into port. One captain assented to the bargain and agreed to hang a burning lantern in his rigging and lead them up the treacherous coast into Provincetown, known to be a congenial harbor for pirates.

But the wily captain had been biding his time. Now the waves were over thirty feet high, the winds were gusting at seventy miles an hour, and Bellamy's men were fall-down drunk.

The captain coasted his small ship over the outer bar and ordered his crew to douse the lantern and throw a barrel of burning tar overboard. The ruse worked. Bellamy faithfully followed the bobbing beacon into the trap while the captain headed for deeper waters. A giant wave slammed the Whydah's stern into the bar and she started to break apart. The next wave rolled her upside down. The Whydah's cannons broke free from their mountings and smashed down through the deck and into the sea chests. A cascade of gold and silver spilled out onto the ocean floor.

The good people of Eastham couldn't believe their eyes the following morning. Over a hundred mutilated pirate corpses lay moldering in the wrack. The beach glittered with gold, silver, and indigo; payment for the seven hundred slaves the Whydah had sold in the Caribbean.

Governor Shute sent the cartographer Cyprian Southack to Eastham to stop the looting, but by the time his boat arrived, more than two hundred men from as far away as modern-day Orleans and Chatham were shoveling gold and rubies into horse-drawn carts and hauling them off the beach.

Captain Southack was only able to retrieve the Whydah's cable and anchor. He issued an irate warning to the citizens of Cape Cod that if they didn't return all the looted treasure they would face the full penalty of the law. The locals merely shrugged and handed back a few beams and some of the ring-studded fingers they had lopped off the pirates' dead hands. For many years, gold doubloons would mysteriously appear in the local Cape Cod economy.

Only two men survived the wreck, the Native American John Julian, who soon vanished into the local Indian population, and the Welshman Thomas Davis. Davis was captured and brought to trial in Boston, where he testified about the exploits of the Whydah before he was hung on the Boston Commons. His body was reputedly left out on Nix's Mate, a small Island in Boston Harbor, as a warning to other pirates.

Though Southack recovered little of the Whydah's legendary booty, he did write up a description of the wreck and note its location on one of his famous maps. Over time the incidents became mostly fanciful stories that Cape Cod grandfathers told to their grandsons to induce them to go to sleep.

One of those young lads was a Martha's Vineyard boy named Barry Clifford. Unlike most boys, who merely brushed off the stories off as simply good yarns, Clifford became obsessed with finding Bellamy's treasure. For nine years he searched through maritime libraries in Greenwich England and at MIT until he tracked down Southack's map, along with his journals and letters. He used these to put together a syndicate of wealthy investors who raised over six million dollars to raise the Whydah.

It is said by some that the real pirates didn't arrive on the Whydah until Clifford discovered her in 1984 and Johnnie Depp did the spirit of Sam Bellamy proud in "Pirates of the Caribbean."

Salt marsh hay drying on staddles. Photographer unknown.

Chapter Seven
Haying the Outer Beach Marsh
August 10, 1720

Delicate blue puffs of sea lavender dotted the shore and the heads of the salt marsh hay were ripe with seed. William Nickerson figured he had just enough time to make one more cutting before the hay collapsed into the long windrows that always reminded him of cowlicks on a bear skin rug.

It had been more than a hundred years since Champlain explored this area. In that time, the slowly rising sea had built up sand spits and shallow coves of water that allowed the salt marsh grass to take root on the backside of the outer beach.

The peat in this young *Spartina* marsh was not very deep. In the older areas of the bay the peat was close to twenty feet thick.

Those marshes had started growing more than four thousand years ago, when the Egyptians were building their pyramids and Hammurabi was codifying his laws in Babylon.

Nickerson purchased the rights to cut the marsh grass from Mattaquason, the sachem of the Monomoyicks. He had paid Mattaquason with a shallop, ten coats, six kettles, twelve axes, twelve hoes, twelve knives, a hat, forty shillings of wampum, and twelve shillings in British coinage. In return, he had received four square miles of upland acreage and the rights to hay the outer beach marsh. It was a fair exchange. Hay was the mainstay of the New England economy. You needed it for food, transportation, even housing. All of Nickerson's neighbors stuffed salt hay straw into the chinks of their homes to act as insulation.

Of course, the best thing about salt marsh hay was that it was free. You didn't have to clear the land or lime and fertilize the soil. The tides did that twice a day for nothing. Many upland farmers preferred to feed salt marsh hay to their cows because it didn't contain the weed seeds that could pass through the cows' multiple guts and come out in their manure to sprout in the farmers' fields of English hay.

For the last few years Nickerson had cut almost twice as much salt marsh hay as English hay. He was even considering transporting his excess to Boston to sell in Haymarket Square. Some of the farmers in nearby Barnstable had been making a healthy profit doing that for years.

Unfortunately, however, Nickerson had not bothered to clear his land transactions with the authorities in Plimouth, and they had been disputing his claim for sixteen long years. Finally, he had to pay a ninety-pound fine and get a written deed from Mattaquason and his son John before the court would grant him clear title to the land.

They had also blocked Nickerson from incorporating Monomoit as a town because it didn't have a large enough population to support a church and minister. It remained as the Constablewicke of Manomoit until 1712, when it was finally incorporated into a town with the far more English sounding name of Chatham. But Nickerson's dubious deed with Mattaquason would continue to plague future outer beach owners for generations to come.

While Nickerson hitched up his oxen to the hay wagon. Walter Eldredge harnessed the horses. Walter was certainly good with horses and had a fine strong back, but Nickerson thought the young buck liked his liquor far too much for comfort. If only he would show up on time for work once in awhile and not get into quite so many fights over local tavern women. The other men had taken to calling Eldredge "Wicker Walter," but the young man seemed to enjoy the sobriquet. This was not a good sign in Nickerson's book.

The farm hands spread out into long lines and swung their scythes slowly, in unison, as they cut their way across the marsh. Walter stayed with the horses on the hard-packed sand at the edge of the bay. The horses wore large wooden horseshoes that looked like snowshoes and were designed to prevent the horses from sinking into the soft peat. More than one draft animal had drowned when it became mired in one of the many mud-filled pannes that pockmarked this thriving young marsh.

When the farm hands were done cutting they drove long slender trunks of black locust into the marsh in large, thirty-foot-wide circles. The peat would prevent the locust poles from rotting, so these staddles could be used to dry the salt marsh hay for years to come.

Nickerson had heard that farmers in the towns north of Boston were starting to build causeways through the marshes, so their horses wouldn't sink in the much thicker peat of the North Shore's older marshes. It took an immense amount of labor. The farmers would remove tons of boulders from their fields and load them into horse-drawn carts, then use them to build the causeways that arched gracefully through the marshes. Nickerson couldn't imagine hefting hundred-pound boulders in ninety-degree weather while fighting off midges, greenheads and mosquitoes.

The causeways were extraordinary pieces of engineering. Extra high tides would cover them with several feet of water, but otherwise they would remain unscathed for centuries to come. Perhaps that is why the North Shore's General Patton knew he could drive his heavy army tanks over the old Roman roads without getting stuck in the soggy Italian countryside during World War II.

By eight p.m., the sun was finally starting to sink behind the cove near Ryder's Farm. Eldredge retrieved his rifle from Nickerson's shallop and drove the horses six miles up the beach to a temporary barn they had built near Pochet Island. He was to spend the night protecting the horses from wolves while the rest of the farmers returned to Chatham in Nickerson's shallop.

It had been a long day. The men took one last draught of water and packed their scythes into Nickerson's boat for the short sail home. They were happy the cutting was almost over. Tomorrow they would toss the hay, so it would dry evenly, and then stack it on the staddles. In a week's time they would pitchfork it into hay carts and transport it back to Chatham. If the weather cooperated Nickerson would have more than enough hay for the winter.

Nickerson thought that next year he might even build a barge so he could sail the hay directly across the bay to his own wharf. On the North Shore, farmers called such barges gundaloes and used them to transport hay across Plum Island Sound.

The following morning Nickerson sailed across the bay, but Eldredge was not on the marsh. He was supposed to have driven the horses back up the beach to meet the men at their predetermined spot. Had he been attacked by wolves or swept off the beach during one of his late night swims? No, he had taken one of Nickerson's best horses and ridden all night long to get off of Cape Cod.

From that day forward, Cape Codders became obsessed with the exploits of Wicked Walter and his multiple descendents. If you ask someone on the outer cape how to tell the difference between an Eldredge spelled with an "e" and an Eldridge spelled with an "i" they have a simple answer, "The Eldredge's with an "e" steal horses, the Eldredges with an "i" do not."

After several successful decades, Chatham farmers stopped haying Morris Island and the outer beach marshes. By then most of the forests had been cut down for firewood, so it was easier to simply grow English hay in your own field, closer to home.

But out on the far reaches of the marsh, the slender trunks of black locust trunks remained buried in the peat, and the ruts made by Nickerson's hay wagons remained embedded in the hard-packed mud, soon to be buried beneath the blowing sands. Like the bones of the sepulchral Sparrowhawk, the sand would preserve them like a perfect fossils until 2008, when a late autumn storm would wash away the overburden of sand and it would look like Wicked Walter had ridden off the beach only the night before.

The beach has rolled over, leaving this strip of peat in the surf. The peat is from the marsh that used to grow in the bay behind the beach. If you look carefully you can see the tracks of the hay wagons and the hoofprints of the horses and oxen that used to pull the carts.

Chapter Eight   39

The Old Harbor Life Saving Station.

Chapter Eight
The 1846 Inlet

Tensions were mounting. Chathamites could sense that the Civil War was just around the corner. They also had a problem closer to home. The Atlantic Ocean was knocking on their parlor door.

The ocean finally burst through Nauset Beach in 1846, creating an inlet for the first time since anyone could remember. A few people knew that Bradford's chart of the Sparrowhawk showed an inlet opposite Strong Island in 1626, and Southack had charted an inlet opposite Minister's Point when the Whydah went down in the 1700's. No one could know that this pattern of a hundred and forty years between inlet formations would prove significant.

The new inlet had created a naming problem, because now there were two beaches; one to the north of Minister's Point and one to the south. Chathamites immediately started referring to them as North Beach and South Beach, dropping the former term,

Nauset Beach, altogether. Only in Orleans did people continue to use the proper nomenclature.

South Beach would break apart much faster than North Beach, most dramatically during a series of storms: the Minot's Ledge Light Gale in 1851, the storm that unearthed the Sparrowhawk in 1863, a hurricane that would wash away two hundred thirty six acres of beach in 1869, and an easterly that finally flattened the remains of the barrier beach in 1871. Only twenty five years earlier, South Beach had been half a mile wide and covered with high dunes. Now in 1871 even moderate-sized waves and high tides could wash over the beach and attack the mainland.

The first place to feel the impact was Chatham's famous twin-towered lighthouse; the second place was Chatham's considerably more infamous Scrabbletown, home of Cape Cod's most proficient mooncussers, known for luring sailors on moonless nights to their deaths.

Prior to the formation of the inlet, Scrabbletown had been a rough but quiet village with stores, gardens and fields. When the inlet opened, it became a raging fury of breaking waves and ocean foam. Scrabbletowners responded by moving their homes into Chatham Center. These Scrabbletown homes would become the smart stores and fancy boutiques of downtown Chatham in the twenty first century.

The best-documented loss from the formation of the inlet was the demise of the Chatham lighthouse. The lighthouse had been ably run by the nation's first female lighthouse keeper, Angeline Nickerson, who had taken over after her husband died at the post ten years before. In 1872 she retired and was replaced by Captain Josiah Hardy.

When Josiah moved his family into the lighthouse keeper's house it was two hundred twenty eight feet from the edge of a fifty-foot-high cliff, but the cliff was now eroding at the rate of thirty feet per year. Five short years later the Hardys could look out their kitchen window and see the ocean crashing against the shore only seventy feet away. In 1875 their daughter was standing on the bank when it collapsed and cascaded down the face of the cliff, carrying the young girl with it. Luckily, she stopped sliding before she reached the surf crashing on the beach below. Fast-thinking lighthouse workers were able to lower a rope to the fortunate girl and haul her back to safety.

By 1879, the lighthouse was abandoned and people simply waited for the inevitable to happen. Boys regularly gathered at the bottom of the cliff to throw rocks at snakes slithering out of the bank as pieces of it fell into the ocean. On December 15, the south tower of the lighthouse leaned over, cracked about three feet above its foundation and pitchpoled down the bank. It landed on the beach below, smashing into a hundred pieces. People from all over town could feel the impact of the crash and hastened to the beach to mourn their loss.

Thirty three years after the inlet had opened the ocean had destroyed two lighthouses, the lighthouse keepers' house, and two streets; over a dozen homes had been moved or demolished. South Beach had migrated over half a mile to attach to the mainland, and North Beach had started growing again.

The Wreck of the Orcutt, December 1896

The final event of that era occurred in 1896. A December nor'easter had stalled off Cape Cod and was lashing at the Great Beach from Provincetown to Monomoy. Clouds of roiling snow and sheets of piercing sleet cut visibility to a scant few feet.

Winifred Nickerson was repairing some gear in his fishing shack on the Chatham shore. Suddenly, the naked masts of a schooner loomed out of a slight break in the squalls. She was hove to just off the outer bar. In an instant, Nickerson knew she was in danger. He had spent five years in the Life Saving Service and knew that the patrolman from the Orleans station had probably missed the distressed schooner on his five-mile patrol to the end of the beach. Now he would be heading north again to arrive back at the station in time for the four p.m. watch.

Nickerson exhorted the small knot of fishermen that had gathered to catch another glimpse of the vessel. "We got to get out theyuh. That vessel will be in ten thousand pieces before mornin'. It may be our only means of savin' life."

"Now take it easy, Winifred. There's time enough to go in the mornin'. If we go over theyuh now, we'll have to stay the night, for we sure as hell can't find our way back after dark in this kind of a storm!"

Much to Nickerson's dismay, the group decided to meet again at seven thirty p.m. to see if the storm hadn't abated enough to try again. Winifred promised that if they were successful, he would walk the five miles up the outer beach to alert the Orleans station.

"Meanwhile, I intend go straight to the Chatham office to see if I can't wire the Orleans Station. Any of you boys willin' to come along?"

Two hours later Nickerson arrived at the Chatham telegraph office, only to discover that the cable that ran underwater to the Orleans Life Saving Station had parted two weeks before. The operator wired the Highland Lighthouse observer in Truro who, in turn, found an open line to Amelia Snow, telephone operator at the Orleans railroad depot. Amelia promised to find someone to hand deliver a message to the superintendent of the Massachusetts Life Saving Service who lived in Orleans. The first person she called was the owner of a local livery stable.

"No, I am sorry, Amelia, but I wouldn't send one of my horses out in such a storm for love nor money."

A young man offered to take the message to Superintendent Sparrow for five dollars, but since nobody was willing to guarantee the money, he demurred and went home. The local expressman said he would gladly deliver the message to Superintendent Sparrow, but he had to wait for the incoming train. It was already

Chapter Eight 43

two hours late because of the storm. Finally, Henry Cummings volunteered to go. Aided only by a sputtering lantern the young Orleans businessman arrived at Sparrow's house at eleven p.m. Benjamin Sparrow rang up the keeper of the Orleans station at once.

"Keepah Charles, this is Superintendent Sparrow. This is rather a wild goose chase, but we have a report of a vessel in distress. We don't know where the vessel is, and we don't even know that she is yet ashore."

"Well, we have to go out but we don't have to return, do we?"

Superintendent Sparrow laughed in spite of himself. Both men knew this was the motto of the United States Life Saving Service.

"I'll break out the beach apparatus as soon as my two surf men return from their patrol, superintendent."

"Good, I'll join you as soon as I make my way out to the station."

"We do appreciate that, sir."

By ten forty p.m. Keeper Doane of the Chatham Life Saving Station started receiving reports from his patrolmen that pieces of wreckage were coming in on the surf. It was a sure sign that a ship was breaking apart in waters to the north. The south patrol found a yawl missing its hull, its bow and centerboard box. Had its occupants been swept overboard as they tried to escape the vessel before she broke into pieces? Further down the beach another patrolman found five hatches and a broken quarterboard with the letters, "Calvin B. Or—". This was even more frustrating. The surf men knew that men were dying just beyond the breakers, but they couldn't reach them, and they couldn't even row across the inlet to alert the Orleans Station.

By the time Keeper Charles' two patrols returned to the Orleans Station at twelve twenty p.m., an eight-foot snowdrift had built up in front of the doors of their boathouse. Seven men had to shovel away the water-soaked snow before they could wheel out the surf boat and hitch it to their single draft horse. For several more hours the phalanx of men had to shovel a path in front of the boat cart and try to locate the faint track that wound through the windswept dunes. They held wooden shingles in front of their faces to protect their eyes from the stinging sand and sleet. Several men suffered frostbite and Superintendent Sparrow lost his vision permanently from the lacerations he received that night. Despite the crew's best efforts, the exhausted horse fell several times and had to be helped to its feet by the equally exhausted men. For the last few miles the surf men were pushing the cart mostly by themselves.

At two twenty five p.m. the crew finally spotted the stricken vessel. She was four and a half miles south of the station and six hundred yards offshore. Her four masts were still standing and anchors were holding her head into the wind, but her hull was already submerged. Breakers were rolling down the length of her decks. Nobody could survive in those waves.

The exhausted team rolled out a small cannon used to fire lines into the rigging so the crew could be winched ashore over the breaking waves in a breeches buoy. But the Lyle gun was useless. The ship was too far offshore and there were no signs of people lashed to the rigging anyway.

The beautiful four-masted schooner that had once plied the oceans from Boston to China was helpless. Wave by wave she was being smashed to pieces on the outer bar. This had been the Calvin B. Orcutt, en route from Portland, Maine, to Virginia, without any cargo. She had hoped to reach Norfolk before Christmas. Instead, all seven of her crewmembers were lost. Frozen bodies washed up in the surf on Monomoy for a month after the ship went to pieces. The sailors had probably fallen from the rigging as they slowly froze to death. Two bodies were never recovered, and two were never identified. Ministers from all the Chatham churches officiated at the funeral of the unidentified men, and their bodies still rest in unmarked graves overlooking the scene of the tragedy.

The wreck of the Calvin B. Orcutt affected history in several ways. The Life Saving Service had been severely criticized for its slow response to the disaster and the crewmembers inability to do anything once they reached the scene. The problem was not the unjustly criticized Orleans crew, but that this part of Nauset Beach was a magnet for wrecks. More vessels passed along this coast during the late 1800's than anywhere else in the world except the English Channel. On a fine summer day, Winifred Nickerson could sit on his front porch and watch over a hundred tall ships, packet boats and fishing smacks sailing just beyond the Chatham bars. At least two other major wrecks, the Onondaga and the Orissa, had grounded within a mile of the same spot the Orcutt had met her demise.

But the most interesting aspect of the Orcutt wreck was where she lay after the storm. She was not on the outer beach but inshore of the 1846 inlet. Records of the wreck stated that pieces of the ship had been swept into the upper beach. But in 1896 the beach was about half a mile offshore from its present location. This piece of the wreck had probably been swept up onto the side of the "old" inlet.

When the Cape Cod Life Saving Stations had been built in 1871 they had been designed so they were eleven miles apart. Patrols would set out from each station then meet and exchange a metal disc at a halfway station before returning back to their own station. There had always been a gap between the Orleans and Chatham stations because of the 1846 inlet. Patrols had to walk to the inlet then turn back. There was no direct communication between the two stations and an even more serious gap in protection.

In 1897, a year after the Orcutt disaster, the Life Saving Service built the Old Harbor Life Saving Station to fill in the gap. At that point the new station was on the southernmost end of Nauset Beach. As the inlet migrated south and the beach continued to grow, the station became a mecca for summer camps, and eventually the dividing point between the north and south camp villages.

Brant hunting. Kristina Lindborg.

Chapter Nine
Unraveling an East Coast Secret
April 19, 1863

It was a cold wet day. Dark clouds hung low in the leaden sky and sleet fell on three men huddled in a cold box on the Monomoy flats. Alonzo Nye had dug the box into the flats in March, then mounded up sand to make a long low sandbar leading up to the hidden blind.

Now Lon tweaked a line attached to several live brant that were eating corn he had scattered in front of the box. Some of his "sports" had been wounded in last year's hunt and Lon had cared for the birds all winter long. Now they were fat, happy and ready for the spring hunt. He tweaked the line again and the live decoys gave their inviting "c-c-cronk, c-c-cronk, c-c-cronk," feeding calls. The wild birds took notice.

Now came the hard part. Alonzo could see that his "sports" were getting more and more excited as hundreds of the large birds swam toward their tiny blind. This was when inexperienced sportsmen usually stood up, shot blindly and scared the entire flock away.

"Hold, it. Hold it. Wait until the first ones climb out onto the sand."

"They'll be right in our laps!"

"That's the idea. Wait, wait."

Finally, Alonzo pulled sharply on the line and the decoys scurried to the right to avoid getting shot. The little traitors to their breed did exactly as they were told.

"Now! Now! Shoot! Shoot!"

Alonzo got forty brant on his first shot; Warren Hapgood and John Phillips shot a satisfying thirty birds each. They had shot three hundred seventy five brant in the last nine days and were well on their way to shooting a thousand birds, a good draw for the month-long hunting season.

The three men were mightily pleased with themselves. Alonzo and Warren had just established the Monomoy Branting Club that had effectively squeezed out the Orleans guides. The club consisted of four resident members who were responsible for the boats, digging in the boxes, making the artificial sandbars and caring for the live decoys. The fourteen non-resident members were mostly Boston friends of Warren Hapgood who were responsible for paying their dues and maintaining the hunting shanty they had just built with their own hands.

The club had been a brilliant idea. Their boxes commanded the best "looks" of the flats. Other guides established the Providence and Manchester Clubs, named rather undiplomatically for the homes of their non-resident members. But the "looks" of the other clubs were never as good as those of the Monomoy Club. In the late 1800's the guides resolved the matter by incorporating all the clubs under one roof, but by that time the sport was changing dramatically.

In 1909, the Commonwealth banned the spring hunt for brant to protect the immature birds. From then on, the club only killed about fifty brant during the autumn migrations when the birds largely bypassed Cape Cod in their eagerness to return to the Carolina eelgrass beds. It was eelgrass that made brant taste almost as good as black ducks, and the spring hunt for the small dark geese was sorely missed. The larger Canada Geese were almost non-existent on Cape Cod in the late 1800's.

The meticulous logs of the Monomoy Branting Club helped unravel one of the East Coast's most mystifying secrets: why the population of brant fluctuated so wildly from year to year.

The club consisted of some keenly observant hunters. Alonzo Nye had grown up exploring the marshes and beaches of Chatham and knew the habits of every species of duck, goose and shorebird that used this flyway.

Warren Hapgood was the president of the Massachusetts Association of Fish and Wildlife, an extremely active organization in that era. Hapgood had corralled so many Boston-based businessmen who knew their lowly place in the hierarchy of the Monomoy Branting Club. Camaraderie, goodwill and considerable ribbing occurred between the resident and non-resident members during their evening-long, after-shooting celebrations.

The shanty certainly beat the clam shack Warren had stayed in on his first visit to Chatham. The walls of the shack had been caked with the dried gurry of recently shucked clams, and the whole, horrible little abode reeked of rotting clam guts. Warren had simply held his nose and buried his head in the seaweed-filled pillow provided for his comfort.

John Phillips was a guest at the club. He taught ornithology at Harvard University and was the world's foremost expert on black ducks. Hapgood had convinced his friend to take the long coach ride from Boston to enjoy this unique form of hunting.

At the end of each day, the hunters would record the temperature, weather conditions, and the number of old and young birds killed. It was easy to tell the young birds because they still had white edging on their coverts and secondary feathers.

The log revealed that in some years their catch would number in the thousands, but in other years it would drop to only a few dozen birds. When this happened the recorder would soberly remark that it looked like brant would never recover, but they always did. The hunters also started noticing that they killed a lot more young birds during the good years.

It was just a few years before the Civil War, and Northerners liked to blame everything on the South. The conventional wisdom was that overhunting in the Carolinas was the cause of the declines. But Phillips used the records of the Monomoy Branting Club to surmise that the drastic declines and recoveries had to originate from conditions above the Arctic Circle. However, meteorological records from the Hudson Bay area were scanty, so he assumed that the problem must be summer storms that destroyed the nesting colonies some summers.

Phillips had it half right. The cause of the precipitous rise and fall of brant did lie in the Northern breeding grounds. But the problem was not summer storms; it was lemmings. The lascivious little rodents would reproduce prolifically every summer until they outstripped the tundra of grass seeds. Then, the entire population of lemmings would be gripped with a suicidal mania. They would swarm across the tundra until they reached the Arctic Ocean where they would jump in and swim until they were exhausted and drown. The surface of the ocean would be covered with their lifeless bodies for miles at a stretch. This happened regularly on a four-year cycle.

The absence of lemmings would reverberate through the food chain. Arctic fox numbers would increase during the summers of lemming abundance. But, after the lemmings committed mass suicide, the foxes moved off the tundra to try to find other sources of abundant food. This left the tundra without one of its most efficient predators, making it possible for the brant populations to swell. If Phillips had had access to the pelt records of the Hudson Bay Company he might have seen the distinct four-year cycle. Today, bird watchers know that in the winters after the lemmings' collapse they are far more likely to see large predatory birds like snowy owl coming south to look for food in New England.

However, when Phillips wrote up his finding in the 1930's, hunters were much more concerned about a fungus than lemmings. The *Labinthulina* fungus had caused a wasting disease that had wiped out the eelgrass beds and, subsequently, the brant that were dependent on them from Cape Cod to North Carolina.

By this time the records of the Monomoy Branting Club had almost come to an end. The Commonwealth of Massachusetts had closed the spring hunt for brant in 1909. The number of brant killed had dropped from an average of a thousand birds shot during the spring hunt to an average of only fifty birds killed during the autumn.

Phillips' records of brant demonstrated the complex ecology of the entire East Coast. Here was a bird dependent not only on the health of unpolluted eelgrass beds from Cape Cod to North Carolina but on lemmings and foxes above the Arctic Circle.

The remains of a shipwreck, most likely the Montclair.

Chapter Ten
"No Book, No Marriage"
Eastham, June 1925

In June of 1925, Henry Beston decided to spend two weeks in the "Fo'castle," a tight little shack he had just built in the dunes two miles south of the Nauset Coast Guard Station. At the end of the two weeks, he found he did not want to leave. "I lingered on, and as the year lengthened into autumn, the beauty and mystery of this earth and the outer sea so possessed and held me that I could not go."

Thus began a classic piece of American literature. But *The Outermost House; A year of Life On the Great Beach of Cape Cod*, would have never been written were it not for a marriage proposal.

For several years Beston spent every morning writing up copious notes about his observations. Reading these, you could feel his senses grow more acute and his meditations more deeply insightful. He was in a zenlike trance of heightened awareness.

But by 1927, Beston had had enough. He was lonely and had to make a living, so he returned to his hometown, Quincy, Massachusetts, and proposed to his long-time sweetheart Elizabeth Coatsworth. But Elizabeth was also a writer and when she read Henry's notes but saw no manuscript she put down her foot, "No book, no marriage." All writers should have such a no-nonsense muse.

The Outermost House was published in October 1928. The couple honeymooned in the "Fo'castle" for two weeks and then hardly ever returned.

The Outermost House is one of the most evocative pieces of writing in all of American literature. It is a deeply meditative discourse on biology, geology and the experience of living alone on the outer beach. But it is not just navel gazing. Beston kept abreast of the latest discoveries in science and made a point of passing all his observations by his colleagues at Harvard, where he had been an undergraduate.

His observation that, "We need another and a wiser and perhaps a more mystical concept of animals," has stood the test of time. He would not be surprised by the recent revelation that giant squid may have drowned ancient Icthyosaurian marine dinosaurs and dragged their bodies down to their underwater lairs where they rearranged their vertebrae into lines so they looked exactly like suckers, self portraits of the giant cephalopod's own tentacles. Beston would have noted that modern octopuses do almost the same thing with their prey. He wrote, "In a world older and more complete than ours, they move finished and complete, gifted with extensions of the senses we have lost or never attained, living by voices we shall never hear. They are not brethren; they are not underlings; they are other nations, caught with ourselves in the net of life and time, fellow prisoners of the splendor and travail of the earth." These observations are still quoted in animal behavior classes and have become guiding principles in the expanding field of ethology.

Even though Beston spent several years in the shack, he wrote his story as a single year in the life of the outer beach. It was a time-honored way to write a book about natural history. It gave Beston's story a beginning, middle, and end, allowing him to write about birth, death, struggle and fulfillment.

He gave his reasons for undertaking the task in the autumn section, "The world to-day is sick to its thin blood for lack of elemental things, for fire before the hands, for water welling from the earth, for air, for the dear earth itself underfoot. In my world of beach and dunes these elemental presences lived and had their being, and under their arch there moved an incomparable pageant of nature and the year."

He wrote knowledgeably about the migrations of birds and philosophically about the hunters who surrounded his little shack in the fall.

He used his many enlightened senses to tell of the arrival of winter.

"There is a new sound on the beach and a greater sound. Slowly and day by day, the surf grows heavier, and down the long miles of the beach, at lonely stations, men hear the coming winter in the roar."

1925 was the stormiest year in half a century. There were five wrecks that winter, including the Montclair that went down with all five crewmembers. Beston not only visited most of the wrecks but was given a haddock by the crew member of a fishing boat that had foundered. He also became close friends with the surf men who visited his shack, swinging their kerosene lanterns on their midnight treks up and down the beach in search of ships in distress.

"Every night in the year, when darkness has fallen on the Cape and the somber thunder of ocean is heard in the pitch pines and the moors, lights are to be seen moving along these fifty miles of sand, some going north, some south, twinkles and points of light solitary and mysterious."

Beston also had a firm grasp of coastal geology. He knew that the outer beach was eroding inexorably landward, and eventually had to relocate his shack to prevent it from being washed away. He has written perhaps the most evocative description of how blowouts form in dunes.

"During a recent winter, a coast guard key post was erected on the peak of the dune; the feet of the night patrols trod down and nicked the crest, and presently this insignificant notch began to 'work' and deepen. It is now eight or nine feet wide and as many deep… The loose sand is whirled up by the wind and poured eastward through this tunnel. At such times the peak smokes like a volcano. The smoke is now a streaming blackish plume, now a this old-ivory wraith, and it billows, eddies and pours out as from a sea Vesuvius."

He wrote of the grinding roar of the surf as well as the plaintive call of the piping plover, "the loveliest musical note of any North Atlantic bird." And he wrote of the night in July when, "The whole night was turned in one strange burning instant into a phantom day. It was from an enormous meteor consuming itself in an effulgence of light west of the zenith."

He could also write pungently, "To my mind we live too completely by the eye. I like a good smell—the smell of the freshly ploughed field, the cloverlike aroma of our wild Cape Cod pinks, the good reek of hot salt grass hay and the high tide blooms in these meadows late on summer afternoons."

He ended his book back in October, "And because I had known this outer and secret world, and been able to live as I had lived, reverence and gratitude greater and deeper than ever, possessed me… CREATION IS HERE AND NOW. So near is man to the creative pageant, so much a part is he of the endless and incredible experiment, that any glimpse he may have will be but the revelation of a moment, a solitary note heard in a symphony thundering through debatable existences of time."

Today when we think of the outer beach, we think of Henry Beston, walking her strand, reading books by kerosene lantern and writing up notes on the Fo'castle's sunlit kitchen table. He set the bar pretty high, not only for nature writers but also for the dwellers of dune shacks who would follow.

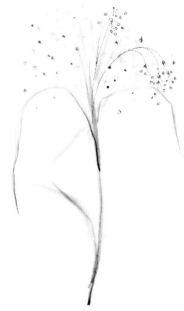
Sedge. Kristina Lindborg.

Chapter Eleven
Rum Runners and Mooncussers
December 28, 1929

December twenty eighth was a clear crisp winter's day. Tendrils of sea smoke rose off the still warmish waters of Woods Hole Harbor, about fifty miles southeast of the outer Cape's Nauset Beach. The low sun shown off the empty mansions of Penzance Point and what the locals called "bankers row" on Church Street. The more modest homes of scientists and shopkeepers clustered around the quiet waters of Eel Pond. Charlie Travers understood that the research schooner the new lab intended to build would house a secret compartment so even the scientists could smuggle their own hooch past the customs officials. "Competition," thought Charlie to himself.

"Evening Captain, looks like a good night for fishin'." It was one of the deckhands of the Coast Guard cutter tied up beside the Black Duck.

"Dunno. Think you'll find any rum runners in all this fog?"

It was a standing joke between the crews of the Coast Guard cutter and the Black Duck, known to be one of the fastest rum runners in Massachusetts waters. Everyone understood that the dock was neutral territory. The coasties knew they couldn't touch the Black Duck as long as she didn't have any liquor on board, and Travers knew the Black Duck was faster than any boat the Coast Guard could afford. All he had to do was steam three miles out to where the supply boats waited, then steam back, unload his cargo and innocently tie up his empty boat beside the Coast Guard cutter the next morning.

Some of the summer people had even built special piers and underground tunnels so that the rum runners could deliver the hooch right into their homes. Charlie remembered the night a crew had climbed up some stairs at the end of a tunnel and emerged into a grand ballroom where a hundred thirsty guests decked in black ties and ball gowns broke into spontaneous applause.

Woods Hole was that sort of town. There was a tolerance for popular things like rum running and speakeasies and a good-natured rivalry between those who broke the unpopular law and those who upheld it.

The only guy who didn't understand that the whole thing was a bit of a game was the Coast Guard's new boatswain. Rumor had it that Boatswain Cornell had quit a higher paying job in the Navy for the adventure of chasing rum runners, and that he had been severely reprimanded by his superiors for shooting up several summer homes and opening fire on a rum running boat in Rhode Island's Jamestown Harbor.

Cornell had stopped the Black Duck on several occasions but

had never been able to catch her with liquor on board. He didn't take kindly to the daily dose of ribbing he had to endure from his happy-go-lucky dock mates. He had even been heard to threaten, "You better watch out or someone might shoot into your wheelhouse." Everyone had laughed at the ridiculous idea.

Charlie turned over his Liberty boat engines and within a few short hours the Black Duck was approaching the rum line of schooners and steamships sitting safely beyond territorial waters. Charlie radioed one of the supply ships just in from Lunenburg that all was safe. There was a rumor flying around among those in the trade that the feds had a new woman in New York who could decode their transmissions and send their locations to the Coast Guard cutters.

Charlie purchased his cargo and headed back toward Narragansett Bay, where his brother was waiting with a truck. He was going full speed as he rounded the Dumpling's Ledge bell buoy, when suddenly the searchlights of a patrol boat pierced through the pea soup fog. Boatswain Cornell had been tipped off about the Black Duck's drop off point, and was waiting in ambush with his lights extinguished. Without giving Travers any warning, Cornell ordered his men to open fire into the unarmed boat.

Charlie felt one of the machine gun bullets shoot off his middle finger, and had to watch as two of his crewmembers and his brother got hit in the chest.

"I radioed back to shore that I had a man all shot up but still alive. The bastards held me offshore for almost an hour and I had to watch as my brother died at my feet. I was still bleeding bad and couldn't wait any longer so I brought the Duck back into the harbor and headed for the Coast Guard cutter. I figured if they were going to finish me off they might as well do it then and there."

The town was incensed at what it considered to be outright murder by Cornell because of his grudge against Captain Travers and the fast boat he could never capture. Reporters were soon swarming the town from as far away as Boston and New York. The incident became a national story highlighting the excessive force used in dealing with those who ran afoul of the hated Vollstead Act. The Black Duck became a rallying cry for anti-prohibitionists and eventually helped lead to the repeal of the unpopular experiment in social engineering. And what happened to the Black Duck? It was requisitioned, repainted and served as one of the Coast Guard's fastest patrol boats during World War II.

The Magnolia incident. Katie McElwain.

One Night in July

The boys met on a moonless night in early July. The beach was dark and empty for miles in either direction. Everyone had brought a lantern to lure in ships and a flask to supply liquid courage for the grisly task ahead.

It was not like the old days when the mooncussers from Chatham's Scrabbletown would mount a lantern on a horse and trot him back and forth through the dunes at night in the dead of winter in an attempt to lure ships onto a dangerous shore for whatever loot lay aboard. In those days any sailor who didn't drown with their wrecked ship could expect to be quickly dispatched by a swift blow to the back of the head. In *Captains Courageous* Rudyard Kipling described those who cussed the moon in more poetic terms, "Ye Scrabbletowners, ye Chatham wreckers, Git out with your brick in your stocking."

Mooncussers instilled fear in the toughest of sailors. The term comes from the fact that sailors could see very well on moonlit nights, and so the decoy lanterns meant to draw ships to shore could not fool any experienced sailor. The would-be thieves were left "cussing the moon."

"Charlie" had learned mooncussing from his "uncle Pete," who had learned it from his father and grandfather. Uncle Pete was the mooncusser who lured the Hedge Jensen ashore the year that Beston was living in the Fo'castle. Many a Cape Cod home had celebrated with one of the Jensen's seven hundred sixty three cases of Cuban rum, and paneled their parlors with fine mahogany from her cargo of sawn lumber. That was the rub with mooncussing; so many people profited from the trade that it was difficult to stamp it out entirely.

While many saw mooncussers as the laziest form of pirate scoundrels, Charlie explained the practice as "a skill, an art, a trade handed down from one generation to the next. But my father got religion after he married so he couldn't be too interested in hearing the stories, at least when my mother was around. By the time she passed away, my father's arthritis had slowed him down so much he couldn't scamper up and down the dunes. He stayed in the used-car business and pretended it was his superior morality rather than his bad knees that kept him respectable. My old man was a great guy but he used to drive poor uncle Pete nuts. 'Self-righteous SOB' was about the kindest thing uncle Pete ever used to call him."

However, the practice declined as more and more ocean-going vessels became equipped with sophisticated arrays of electronic gear that could keep them off the sand.

While it was increasingly difficult to lure ships to shore, there had been one night when they thought they had a customer. Bass River Jimmy was stationed offshore, rowing his dory back and forth, when sure enough he signaled that he saw a north-bound ship sailing close to shore. The boys waved their lanterns around madly until the ship's running lights loomed into view.

"We thought we were in for a big score, when all of a sudden this guy shot a searchlight at us that could have lit up the beach all the way to Portland, Maine. It was like looking into the sun. Next thing, we hear machine gun fire, and then there were flares, and then something that blew away thirty feet of the dune right behind us. In the light of the flares we could see Jimmy rowing that dory, which must have weighed nearly a ton, like he was pulling a scull in the Charles River regatta. I wouldn't have believed the boat could move that fast.

"Then the SOB heaves to about a quarter of a mile off the beach and keeps shooting in our direction. One of the boys swore that he heard laughing coming from the ship. We figured that if he wanted to, he could have wiped us all out, but we were too scared to move. Finally, they let off one more round of whatever the hell it was—we thought the dune was going to collapse—and headed back northwards. We never did find out what kind of ship it was; we were too shaken up to ask a lot of questions."

Charlie continued, with a story of the boys last outing. "This one night was different. It wasn't long before we were pretty well oiled and we started to wave our lanterns like crazy to attract a ship. The weather had turned pretty foul and all of a sudden we could see both the red and green running lights of a large vessel looming through the murk. She was heading straight for shore.

"The adrenalin started pumping through our tired old bodies. We were mooncussers, goddammit, and here was a customer heading right in to greet us. The running lights got closer and closer and then there was the long scraping sound of the ship's keel running into the sand. The visibility was lousy but we could tell that it was pretty big. Booty, plunder, flotsam and jetsam, we were eager for the kill.

"Once she settled on the beach we ran over to see what we had captured. But as we got close we noticed an awful stench coming from her direction. It was like all the dumps on Cape Cod steaming in the sun on the hottest day of the summer.

"One of the boys remembered reading an article about a New York garbage scow that couldn't find a home. He had been amused because the scow was called the Magnolia. We couldn't see much with the lanterns, but fortunately Bert Ellis had a nice powerful flashlight. It only took two seconds to find the name on the barge's bow. There it was in big white letters, 'Magnolia.' The barge known 'round the world,' the best known piece of maritime equipment since the Titanic. And, sure enough, we had mooncussed the goddam thing right onto Nauset Beach. We didn't know whether to laugh or cry … it could have gone either way for most of us. Well, we headed off, briskly, I should add, going our own separate ways, thinking our own separate thoughts."

Bass River Jimmy retired to Florida to escape the eternal scorn of his fellow mooncussers and "Charlie" ended up as a widower in a nice quiet nursing home in Brewster. It was the end of the road for the last of the mooncussers on Cape Cod.

The distinctive Indian club shape of the end of Monomoy Island. Notice where re-curving spits have encompassed salt water that later became fresh water ponds. Photo, courtesy of US Fish and Wildlife Service.

Chapter Twelve
Monomoy Island
August 7, 1958

Our government-issued jeep is loaded with clam rakes and fishing rods, olive oil and kerosene—the essential necessities for my family's annual cooking and fishing expedition to Monomoy Island on the southernmost tip of Cape Cod. We hold our breath as my father inches our jeep up two narrow planks and onto the rickety barge that serves as the lone ferry to the nearby island. A jeep is the only vehicle that can fit on this homemade platform. The barge owner wraps a piece of old fraying cord around the starting wheel and pulls. Nothing. He curses and pulls again. Swearing seems to help; the motor pops and sputters into life and we move slowly over the shallow waters.

Just before landing we see the rusty hulks of several old Model T's mired in the sand. My father eases the jeep back down the narrow planks and we deflate the tires to avoid a similar fate. Slowly, we lurch and chug down the sandy ruts that run the length of this eight-mile-long island. The greatest thrill for my sisters and me is to sit on the tailgate where we can feel the sand spinning off the tires and the heat of the exhaust singeing our naked legs. The jeep moves so slowly we can jump off, land on our backsides, run a few steps and still have enough time to leap back on the moving tailgate. We are told to stop our shenanigans but continue to pretend to be thrown off on every bounce, while gales of laughter give us away. Other times we are told to jump out and shove boards under the tires to give the jeep traction through the sandy dunes.

It is easier once we make it to firm sands of the foreshore. Here all we have to worry about is the oncoming breakers that always seem larger than those on the mainland. We swerve up and down the beach to avoid them, and it is dark before we reach the far end of Monomoy. Occasionally, we see the headlights of another beach buggy lurching back and forth through the dunes. But mostly we are alone, save for the ghostly apparitions of beach toads trying to scramble out of the ruts as the bouncing beam of our headlights bear down upon them.

The tracks on the backside of the island are as firm and smooth as asphalt. Flocks of shorebirds huddle in the marsh grass, and horseshoe crabs mate along the shore. Finally we see the low-lying silhouettes of a handful of lonely dark shacks huddled around the sandy cove called the Powder Hole. Our shack is a casual affair with an outhouse, kerosene lanterns and thin wooden walls. It is my job to prime the hand pump with a bucketful of water left by a thoughtful former occupant. Soon, sweet fresh water is gushing into the tin basin. After the kerosene is poured and the lanterns are lit we spread our sleeping bags onto simple wooden bunks and fall into a deep slumber. The only light is the long beam of the Monomoy Lighthouse that sweeps over the dunes, into our windows, and out again over the black Atlantic.

We wake up early to be on the point before sunrise. Monomoy Point is a place of shoals and sandbars, rips and runnels. This is where cold green waters of the Labrador Current mix with the warm waters of the nearby Gulf Stream. This is where the twelve-foot tides of the North Atlantic flow into the three-foot tides of the mid-Atlantic. This is where the combined energy of wind and water delivers millions of tons of sand that has eroded off thirty miles of continuous beach and bluffs from Truro to Chatham. Here the sand piles up in treacherous shoals and sandbars. Here the sandbars migrate onto the beach where the wind fashions them back into thirty-foot sand dunes. The dune lands stretch almost two miles thick from the ocean to the backside, making this southernmost tip of Monomoy look like the head of a giant Indian war club that connects to a long narrow handle.

We cast our jigs into the maelstrom and bounce them back along the bottom. The currents are so strong we can't reel them in against the tide. A large striped bass lunges out of a nearby hole, and terns gather raucously overhead. We can see bluefish bursting through the green wall of a current to attack a school of squid. The water turns reddish black with the squids' ejected ink. Seals gather to harass the bluefish, and where there are seals, sharks may not be far behind. Our neighbor swears he once saw two great white sharks continuously beaching themselves on Monomoy to trap menhaden against the shore. After each lunge the sharks would have to flip around like giant minnows to make it back into the water.

The bluefish come in fast and furious, until the tide slacks and we change bait to catch some fluke. After awhile, I walk into the dunes to nap in a patch of sweet-smelling beach peas. By late afternoon we return to the Powder Hole to dig some clams and watch my father clean our fish on a simple wooden cleaning table built into the side of the shack.

We hear my mother pouring olive oil into a hot frying pan and run inside to help her dip the fillets into cornmeal and watch their edges curl as they hit the hot oil. Soon the entire shack fills with the smell of steamed clams, sizzling fish, hot butter and cold beer.

After dinner, conversation drifts toward the idea of creating the Cape Cod National Seashore. My father is in a position to do something about it. He has been executive director of a national commission set up to look at the future of the National Park Service.

One problem the commission has unearthed is that while most national parks are in the West, the people who want to use them are mostly in the East. The East Coast city dwellers that can afford it flock to places like Cape Cod, Assateague Island and the Jersey Shore—places that are already developed and getting more and more crowded.

So the commission created a new category of national park called a national seashore. This new category gave the Department of Interior greater flexibility in buying land in already-populated areas in order to create the new seashores.

Beach grass.

Chapter Thirteen
The Debate
February 1973

The late Sixties and early Seventies were an extraordinary time of environmental awakening. Books meant something, people listened to scientists, and there was a bipartisan willingness to do the right thing.

The era was ushered in by the publication of Rachel Carson's classic, *Silent Spring*, which is said to have been one of the most eagerly awaited and equally dreaded books to ever roll off a press. Carson had already attained a much-deserved reputation as an elegant writer about nature. Her books *The Sea Around Us*, and *The Edge of the Sea* had been on the top of the *New York Times* bestseller list for several months and had made her a household name.

Dr. Carson was also a well-trained biologist who worked for the National Fisheries Service in Washington D.C. This job gave her access to numerous scientists who usually never talked to each other. Yet all of them had started seeing environmental problems within their own narrow fields; ornithologists were seeing a decline in some species' of birds; ichthyologists observed a decline in fish in some bodies of water, and a chemists noticed that chemicals were building up in the systems of numerous species of animals. Carson's genius was to link all these disparate findings together into a coherent pattern and then make a forceful argument that pesticides were killing birds, fish, mammals and even the humans who handled them. No wonder the chemical industry dreaded the publication of her book.

Silent Spring was an huge wakeup call. Suddenly, people realized that air, water and land were being polluted at an increasingly rapid rate. Rivers like the Cayuga in Cleveland were so polluted they would catch on fire and burn for several months. The Appalachian Mountain forests were dying from acid rain, and the oceans were being choked with tar balls and oil from shoddy shipping practices.

Environmental scientists, who had formerly led quiet academic lives, suddenly found themselves being sought out by journalists and the public at large. They led informed debates and helped engender a bipartisan willingness to discover the best ways to cope with the many facets of the environmental crisis. Gradually, an interwoven network of laws and legislation was created to protect the environment.

One of the questions that arose during this time was what to do about the many species that were perilously close to being exterminated by hunting and the use of dangerous pesticides. America's history in this regard was not good. Passenger pigeons that used to darken the skies and fill larders during Colonial times were wiped off the face of the earth in a few short generations. Seemingly endless herds of bison that had supported scores of Plains Indian tribes were down to just a few thousand individuals.

Whooping cranes, whose elegant courtship dances graced the prairies, declined in number to a paltry sixteen nesting pairs. Even the majestic symbol of our country, the bald eagle, was allowed to dwindle until it was teetering on the very edge of extinction.

Carson's book help spur the public, who pushed Congress to pass the Endangered Species Preservation Act of 1966. It provided fifteen million dollars, a meager sum, to buy land known to support endangered species.

A remarkable group of scientists and policy makers realized this was not going to be enough. They centered around Russell Train, the ultimate Washington insider. Train had attended the toney St. Albans School in Washington and then Princeton and Columbia law schools. President Nixon had first noticed this liberal Republican for his ability to raise campaign contributions for Republican candidates, then for his work establishing several national environmental organizations, including the World Wildlife Fund and The Conservation Foundation. The president appointed Train as the first head of his newly formed Council on Environmental Quality. He was joined by Gerald Bertrand, a marine biologist pulled in from the Army Corps of Engineers, and David Challinor and Lee Talbot, two scientific administrators at the Smithsonian Institution. Together, this small group of committed people convinced President Nixon's domestic advisor, John Ehrlichman, that improving the Endangered Species Preservation Act would make good political sense.

"Nixon's decision to get behind improving the 1966 Act was the turning point for us," remembered Bertrand. "It allowed us to be aggressive and incorporate dozens of new principles and ideas into what would become a landmark piece of legislation. We were helped by the fact that developers were not on their game. They didn't have the cadres of lobbyists they have today that can stymie a piece of legislation before it has a chance even to be conceived. Our work didn't have anything to do with politics, it was just a group of well-informed policy makers and scientists trying to craft the best way of dealing with a problem."

In 1973 Nixon signed into law the completely rewritten Endangered Species Act. This newly-strengthened law changed the direction of conservation in the United States. The California historian Kevin Starr called it the Magna Carta of the environmental movement. Unlike its predecessor, the new act had teeth and was specific. The discovery of a single endangered species in an area could stop development in its tracks. There were no gray areas where experts could disagree about the importance and costs and extent of a particular piece of habitat.

From its inception, the act had broad bipartisan support. One of its co-sponsors was New York Senator James Buckley, brother of the conservative icon William F. Buckley. Another co-sponsor was the liberal Republican Congressman from Rhode Island, Lincoln Chaffee.

It would probably astound today's ardent anti-environmentalists to know that this and many other pieces of legislation that underlie our present system of protecting the environment were conceived of and signed into law during President Nixon's one-and-a-half administrations. Even though many of the bills stuck in the craw of the major chemical and energy industries that supported him, President Nixon was a realistic politician who knew when to bend to the will of the people. In this instance at least, he deserves recognition as the first modern president since Teddy Roosevelt to do so much to create the laws that protect our environment today.

But problems arose almost immediately with implementation of the new Endangered Species Act. Developers suddenly realized that their projects could be thwarted, and they became adept at what became known as the "shoot, shovel, and shut up" policies for evading the reach of the new law. When that didn't work, they pushed Congress to step in and enact special amendments to circumvent the bill. One such amendment allowed the Tellico Dam to be built after it had been initially blocked by a three-inch-long endangered species of fish called the snail darter.

The protection of ugly, unknown species was just too much for some. President Reagan's environmental appointees tried to gut funding for the act, and subsequent administrations have cut the budget of the Endangered Species Office, so that today it is impossibly slow and difficult to add new animals to the endangered species list.

One of the central debates that surfaced during the crafting of the bill was whether to emphasize conservation of endangered species or conservation of the habitat in which they lived. A habitat that supports an endangered species is sometimes obvious and discrete, like an island, a barrier beach, or a specific body of water. But often they can be much more difficult to characterize.

Scientists studying acid rain discovered that you needed to know how all the rainwater and minerals came into a forest and eventually flowed out of a watershed before you could explain how acid rain damages a forest. They found, for instance, that a watershed that lay on the dry side of a mountain could have an entirely different biochemical system from the watershed on the rainy side of the mountain. So, while at first blush it might seem more sensible to protect an entire mountain, scientifically it could be argued that protecting only the watershed that harbored an endangered species was necessary.

There was also the all-important question of public perception. Many environmentalists argued that the public would identify more readily with a fluffy little endangered species like a piping plover than a seemingly barren habitat like a barrier beach.

In the end, the fluffy soft environmentalists won out over the hardheaded scientific environmentalists, and Congress passed an act that often supports putting large amounts of time and energy into saving a handful of species at the expense of people, their pleasures and passions.

This approach has created major resentment towards the very species that the act was established to protect, from the spotted owl in the Pacific Northwest to the piping plover on the barrier beaches of the East Coast.

At the same time that the Endangered Species Act was wending it's way through the Congressional meat grinder, a lesser-known and far less controversial bill was also being passed. The Coastal Zone Management Act was designed to encourage states to create management plans for their coastal areas. Unlike the Endangered Species Act, this act emphasized the importance of the habitat and its resources as a whole over any particular endangered species.

As we are becoming more and more aware of the consequences of sea level rise, many coastal advocates now think we would be better off if the conservation emphasis was on protecting systems over individual species.

It would certainly be more palatable in today's anti-environmental climate if communities managed their beaches under the guidance of the state's Coastal Zone Management office rather than under the stricter mandates of the more controversial Endangered Species Act. But at this point in time, changing horses in midstream would cause a major upheaval in environmental law and legislation. And if we were to open up that particular can of worms we might run the risk of having Congress throw the baby out with the bath water.

The nude beach. Courtesy of The Provincetown Banner *newspaper.*

Chapter Fourteen
The Nude Beach Battle
Truro, Agust 25, 1974

The early 1970's was a time of energy and drastic change. The Vietnam War was winding down. President Nixon had just resigned and the Middle East oil embargo had sent gasoline prices soaring.

On Cape Cod, it was a time of expanded field research. Ian Nisbet was a studying terns for the Massachusetts Audubon Society. The Cape Cod National Seashore had established a research division at UMass Amherst. George Buckley was studying horseshoe crabs on Pleasant Bay, and I was in exile in at the Law of the Sea Conference in Caracas.

Lord, how those delegates droned on and on in mind-numbing detail about territorial waters, manganese nodules and international straits. During one of the plenary sessions a friend from the Sierra Club passed me a copy of *Newsweek* magazine. Next to the small article about an outbreak of spinal meningitis was a picture of hundreds of naked students on Truro's Ballston Beach. I had been studying how horseshoe crab blood could be used to diagnose meningitis, and Ballston Beach looked like another Woodstock, only with more drugs and considerably fewer clothes. Seemed like a good time to get back to doing field research. I booked the next flight to Cape Cod.

As I drove up to North Truro I was not disappointed. Hundreds of cars were parked on the sides of the road. People were getting out of their cars, taking off their clothes and climbing down the dunes to the beach below. One woman was doing elaborate yogic poses as she slowly made her way down the dune. It took her half an hour to reach the bottom.

Rangers from The Cape Cod National Seashore were also taking notice. They didn't particularly like being touted as having about the best nudist beach in the United States. Superintendent Tom Hadley wanted to get some scientific evidence to back up a ban on nudity. He asked Paul Godfrey, the head of the National Park Service's Research Division, and his assistant, the geologist Steve Leatherman, to produce a paper that would show how much nudists were damaging the dunes. The whole idea seemed underhanded to Steve Leatherman. Science shouldn't be used to advance a political agenda.

The first time Steve had come across nudists was when he was walking along the beach below Highland Light. It was a place where blue clay oozed out of the bottom of the cliffs and formed inviting sun-warmed clay pits.

Steve wasn't paying much attention, "I was just looking at the shells and enjoying myself when all of a sudden twenty nude blue people jumped out of the pits and plunged, yelling and screaming into the ocean. It was pretty bizarre."

Steve figured that if people wanted to enjoy the beach without any clothes they were perfectly within their rights. If parking was a problem you should just tow a few cars for a hundred dollars a pop, and the situation would quickly resolve itself. But Superintendent Tom Hadley didn't see it that way. He wanted scientific proof that nudies were trampling the dunes.

Steve's research showed that the entire outer arm of Cape Cod was eroding back at rate of two to feet a year and these cliffs sometimes lost twenty feet in a single storm. Even Thoreau had written about seeing a sixty-foot hunk of the Highland Light cliff collapse and plunge to the beach below. What little damage the nudists could do during a few weeks in the summer was small potatoes compared to what nature did every year.

But Superintendent Hadley wanted the data, so he would give the superintendent data to show that nudists were causing erosion. It was just another contract.

"I hate to say it, but sometimes superintendents of seashores want to use science to support their own political decisions. Paul and I had some words about the role of science, but in the end we wrote the darn paper. I never told Paul that every evening I would take my graduate students to the cliffs and we would coast down the dunes on our butts to swim in the ocean. It was lots of fun!"

Their paper was used to back up the only anti-nudity ban in a national park. Of course, this fell right into the hands of eager journalists. Who didn't want an all-expenses-paid weekend on a nude beach on Cape Cod? The *New York Times*, *Time* magazine and *Newsweek* all featured the controversy in numerous articles.

In 1975 the Massachusetts Civil Liberties Union filed a lawsuit against the ban. Alan Dershowitz, a Harvard law professor and Truro summer renter, and Jeanne Bell, also a faculty member at Harvard Law School, argued the case. In the afternoons you could see them standing buck naked in the ocean, preparing their briefs. They could have been having tea in the Harvard Faculty

Club if it hadn't been for their lack of clothes.

Dershowitz told a reporter that if there were black people congregating on that beach, no one would dare ban them. The powers that be should just solve the parking problem. The court didn't buy his argument.

In the end the federal judges ruled that there was no fundamental right to bathe nude at Brush Hollow Beach, and that the government had legitimate traffic and environmental concerns.

The Cape Cod National Seashore is still the only national park in the country with an anti-nudity regulation. Rangers still write out hundred dollar citations, but they have dropped to about eighty a summer, with a much higher number of warnings. Today nude beach goers are accustomed to covering up when a ranger drives down the beach. Judging from the warning the rangers issued to my former wife, the rangers certainly seem to enjoy their anti-nudity patrols.

Chapter Fifteen
Field Research
Truro, June 20, 1977

It was early morning in the Pamat River Valley of North Truro. A giant snapping turtle floated in the shallow waters of the strange little river that meandered through thick stands of cattails as it made its way across the narrow wrist of Cape Cod, only a few feet above sea level.

Steve Leatherman wondered how such a large fresh water animal could ever get from distant ponds to this isolated river. You often saw the females walking on land after laying their eggs, but it would be quite a feat to walk to the nearest fresh water pond from here. It was even more curious how the small fresh water fish got into these isolated Cape Cod ponds.

He would have to ask his friend Skip Lazell the next time the herpetologist dropped by the lab. Skip always brought along a bag full of fat writhing black snakes for dinner. It tended to get the grad students' attention. Steve always made a big point of asking Skip if there were any snakes he wouldn't eat, "Just the small ones, they don't have enough meat on 'em."

Steve once visited Skip at a bird rehabilitation program on the North Shore. An older woman came in carrying a large fat wild duck she called 'Herbie' under her arm. While Skip took 'Herbie' into the back room to fix his wing, Steve stayed in the front room calming the nice old woman down. Finally she was ready to go, but asked if she could see 'Herbie' one last time. Steve went into the back room and found that Skip had already killed and cleaned the duck and was plucking its feathers for dinner. "Skip, what the hell are you doing? That woman wants to see 'Herbie' one last time."

"Tell her 'Herbie' is having a setback and it would be better if she didn't see him right now."

Funny memories, thought Steve as he unlocked the door to one of the old lighthouse buildings that the Cape Cod National Seashore had converted into a dry lab. He was already making notations on a large pad of white paper as his Earthwatch students straggled in to make themselves coffee.

Steve was quickly becoming one of the world's experts on the migration of barrier beach islands. He had done his thesis in Asseteague, Maryland. A hurricane had broken through Assateague Beach in 1933, creating an erosional hot spot. The new barrier beach island was cut off from its supply of upstream sand by a series of jetties, so the front of the beach was eroding ten to twenty feet every year and storms were washing that sand over the beach and into the marsh. In essence, the beach was rolling over itself as it migrated shoreward.

Before his work, experts thought that beaches either eroded or accreted sand, but were essentially stable. Steve had shown that barrier beaches were dynamic geological features that pulsed and moved like a living organism as they migrated ever landward under the influence of rising seas. Today he was in charge of an Earthwatch expedition comprised of paying participants and led by graduate students. The first team was looking at the effects of ORV's on animals that live in the marsh. The results of their studies would be used to decide whether ORV's should be allowed to continue driving on barrier beaches.

"OK, lets get started. Nancy, how about your team? I understand you had some kind of trouble with your clams."

"Trouble? Oh no. Nothing like that, Except, yesterday we planted a thousand clams on the flats and today only two of them are left."

"You mean I'm going to have to go back to UMass and explain to the comptroller that I have to order another ten bushels of soft-shelled clams from Maine? He is already asking me about the bill for beer and research lobsters. So, what really happened out there?"

"Well, we didn't realize that there was a flock of seagulls watching our every move. After we left, the little thieves swooped in and ate every last clam."

"Maybe you should change your thesis topic to: the spatial memory of seagulls."

The grad students all had a laugh at Nancy's expense. Their Earthwatch volunteers weren't quite sure what to think. Was this the way field research always worked?

"Well, next time you are just going to have to stay out there until the tide covers the flats. Other than that, how does it look?"

"We planted two rows of clams and protected one with a fence. The other is unprotected so ORV's can drive over them. I expect the clams will be crushed even though they are several inches under the sand."

"OK, but I think you will have to drive over them yourself. There aren't enough ORV's driving around randomly out there to give you sufficient data."

"I was afraid you were going to say that."

"OK, good job. Wheeler, Zaremba, what about your project?"

"I'm looking at how vegetation responds when it is covered by overwashes of sand during storms. Last month we dumped a fifty-five-gallon drum of sand over some dune grass, burying the plants under four feet of loose sediment. Frankly, I thought we had killed the grass for sure, but yesterday we saw the first shoots sprouting back up through the sand."

"Impressive stuff. I guess that's why they call it *Ammophila*, 'the sand lover,' but the thing that can kill *Ammophila* is when sand is removed from its roots. It is very sensitive to that. That's why you don't want to have a bunch of footpaths going through the dunes. When a person steps beside *Ammophila*, sand grains slide into the footprint and eventually the beach grass will die.

"Well, I guess that's it. I'll head out to Race Point to see Brad Blodgett's work then join my own team in the Provincelands.

Steve's team was trying to determine the geological and ecological history of the Provincelands. They were a large expanse of rolling dunes pockmarked with swales of wild cranberries. This was where Cape Cod cranberry industry got its early start. But his team was also having problems. Their samples were showing that the swales were underlain with strata of wind-blown sand. Such Aeolian sand meant that the swales had not always been there, so their vegetative samples couldn't really tell them much about how the complex area had been formed. They would have to use a different technique next summer.

The problem with Steve's work at the moment was that the

swales were full of poison ivy. He had always been an outdoors kid, but ever since he could remember he'd catch poison ivy by just looking at it.

It was a difficult problem to solve where he grew up in Charlotte, North Carolina, but it was easy to solve here on Cape Cod. The temperature often rose to over ninety degrees in the breathless swales, but in the evenings the team could slide down the dunes and plunge into the cool green waters of the Atlantic. Salt water was just the thing to remove the poison ivy's itchy Urushiolic resins.

Steve always gave informal symposia after supper. He particularly liked to regale his students with stories about his work trying to reconstruct the history of the Outer Beach. He told them how he had discovered Southack's chart of the Cape that showed the location of the Whydah in 1717, explaining that the wreck was only a thousand feet offshore from where they swam every evening.

His students snickered at his claims that the pirate Sam Bellamy fired his cannons at the top of the Wellfleet cliffs as he was going down. "Yeah right, pirates off Wellfleet Beach! Was he signaling Maria Hallett that he was coming home, or was he trying to ward off any mooncussers who might want to steal all his gold doubloons?"

"Those were good times. I've always said that using Southack's charts to locate the Whydah was one of my greatest academic achievements and one of my greatest economic failures. I never knew that his ships were laden with gold from the Spanish Main."

Chapter Sixteen
Coast Guard Beach
February 5-8, 1978

Prior to the creation of the Cape Cod National Seashore, my father and a few sportsmen like him used to drive a motley assortment of Model T's and woodies down the beach to reach their favorite fishing spots. But even diehard beach buggy enthusiasts knew that beaches were not really appropriate places for automobiles, and realized that too many vehicles driven irresponsibly could destroy the dunes and interfere with the natural processes that allowed beaches to act as natural barriers to protect the mainland.

In fact, organizations like The Massachusetts Beach Buggy Association were created up and down America's coasts to help police its members and support the creation of national seashores. When the seashores were established, they grandfathered in the use of beach buggies, with the understanding that such vehicles would eventually be banned from public lands.

The superintendent of the first national seashore in Assateague, Virginia, said it best, "In brief, the staff here recognizes that beach vehicles are destined to be banned from the public beaches… It is just a matter of time before the outcry against them becomes stronger than the political pressure exerted by them."

But that turned out not to be the case. Detroit realized that they could make a bundle of money if they started marketing a new class of vehicles called Off Road Vehicles, or ORV's, and later, Sports Utility Vehicles, or SUV's. They were mostly aimed at wealthy suburbanites who could drive them back and forth to their city jobs, and then use them for "the ultimate escape" on weekends. Slick television ads showed them roaring up snowy mountainside trails or skidding in long, slow-motion circles on deserted beaches.

Before these gas guzzlers came on the scene, you needed a shovel and a certain amount of expertise to drive a government issue jeep down the beach and dig yourself out when you got stuck in the sand. But, the advent of the new ORV's and SUV's meant that any weekend warrior could now cruise up and down in the new national seashores and sleep in self-contained campers in overcrowded bull pens on the most popular beach spots.

The new vehicles also made it easier for those few people fortunate enough to own or lease camps in the Cape Cod National Seashore to spend time in them in the summer. Small villages of camps became established, and developed their own unique characters. The image of Henry Beston writing by lantern light in Eastham was the standard. The dune shacks of the Peaked Hill area near Provincetown also housed painters and writers like Eugene O'Neil and Harry Kemp.

A massive act of nature was about to change all that.

The Blizzard of 1978

On February fifth a mass of cold Arctic air bulged out of Canada, swept across the American continent, and merged with warm air spiraling off the coast of Cape Hatteras. This created an extra-tropical cyclone with an intense low-pressure core that swept rapidly north.

Weather forecasting had improved dramatically during the early Seventies but this kind of winter storm was still very difficult to predict. So, when the storm failed to arrive early Monday morning, most new Englanders thought it was just another lousy forecast and went about their business. This left them with little time and no inclination to prepare for a blizzard that would eventually kill a hundred people, close down all Massachusetts roads for a week, and cause one billion, seven hundred fifty million dollars in damage.

Most of the casualties occurred in coastal communities, where hundreds of homes were destroyed and almost six thousand were damaged or removed. Most importantly, the storm started to change how we think about protecting our coasts, which, in turn, led to new laws and regulations.

The storm arrived in full force on Monday at about ten a.m. It wasn't long before people realized this could be the storm of the century. By one p.m. snow was drifting on the streets and winds were blowing at forty miles per hour. Most schools and offices started releasing workers at two p.m. Disabled cars were soon blocking both city streets and major highways.

Many people were stranded overnight in their cars on Interstate 95; some were overcome by carbon monoxide and their frozen bodies were discovered in the following days. Others were finally rescued by cross country skiers and snowmobilers. More than three thousand, five hundred cars were left abandoned on the highway during the storm and had to be removed before crews could plow away the huge drifts of snow.

It was standing room only on the last train returning to Boston from New York City, and it looked like it was pulling into Vladivostok, Russia, when it finally arrived in Boston's steamy South Station. The station was packed with people who had missed their flights and had to console themselves with the prospect of a week of cross-country skiing in Boston.

The real problems started on Monday evening when the new moon tide was at its crest. In Chatham, near the center of the storm, the winds were clocked in at ninety-two miles an hour. These hurricane-force winds, combined with the extremely high tide and the intense low pressure area created a huge storm surge that flattened barrier beaches, flooded coastal communities and stranded thousands of homeowners as the seas crashed over ten-foot sea walls. These conditions persisted through four consecutive tidal cycles, compounding an already severe situation.

The entire five-man crew of the pilot boat Can Do was lost when they tried to come to the aid of a Coast Guard patrol boat in Salem Sound. Over ten thousand people were evacuated from coastal communities, including patients in nursing homes that had to be rescued in small leaky boats.

The Old Orchard Amusement Pier was swept out to sea along with Rockport's famous Motif number one fishing shack. Boston's Pier 4 restaurant lost its banquet hall when the Peter Stuyvesant, a paddle wheeler, tied to the dock, capsized and sank in the windswept harbor.

On Cape Cod, the blizzard didn't seem so severe at first, only dropping eight to twelve inches of snow. Coastal dwellers could remember feeling the warmer subtropical wind blowing in off the ocean. This quickly turned the snow into rain and after the center of the storm passed on Tuesday the sun came out and it turned into a seasonal but blustery day.

But the winds continued through several tidal cycles. North Beach in Chatham was completely covered with water and the barrier beach broke through in several places. Almost every camp was lifted off of its foundation and moved hundreds of feet downstream. Debris from the camps littered the shore. Monomoy was severed in two, creating North and South Monomoy Islands, which would last for over thirty years.

The most far-reaching damage occurred in Eastham. Prior to the blizzard, people had assumed that driving across the beach in this area did no harm. But actually, the weight of many cars repeatedly driving across the beach prevented the dunes from forming; without the dunes, erosion was left unchecked.

When the sun came out on Tuesday morning people flocked to the ragged edge of Eastham's Coast Guard Station beach parking lot to see the situation. The storm surge was sweeping all the camps out to sea or up along the marshes of the mainland. The entire parking lot was damaged and would have to be moved back inland.

For many, the greatest tragedy of all was the loss of Henry Beston's beloved Fo'Castle dune shack. It floated off its foundation Monday night and split into several pieces. Some swept through the inlet and out to sea, while others were scattered across the beaches and marshes of Eastham. People flocked to the shore to get souvenirs of the literary monument. By the end of the storm, only three camps were left standing. These were gradually abandoned during the next two years.

The town of Eastham decided to close Coast Guard Beach to further ORV traffic. According to Eastham's Natural Resources Director Henry Lind, the results were dramatic. "Now, you could stand at the Coast Guard Station facing south and see a very well-established and healthy beach and dune system." The decision to ban ORV's from this section of the national seashore would reverberate up and down the East Coast, causing ORV enthusiasts to fight back. Their demands for increased beach access would become more and more strident.

A piping plover. Kristina Lindborg.

Chapter Seventeen
Piping Plover
Nauset Beach, May 15, 1986

Eric Strauss and Laurie MacIvor drove slowly down Nauset Beach. The wind off the early spring ocean still held a chill, so the researchers had to look carefully for newly arrived piping plovers hiding in ORV tracks to stay warm.

Piping plover have been described as ridiculously cute, like the cotton candy chicks you buy at Easter. Their little toothpick legs are a blur when they run in and out of the waves before stopping abruptly to stare up at you and give their plaintive little "pweep, pweep" calls. If you are sunning yourself in their territory they will pay you no mind. They might even run across your beach towel.

However, if you are another piping plover, watch out. Piping plover are capable of pecking to death any plover who horns in on their territory. Plover couples defend a ten-foot-wide corridor from their nest to the shore, where they spend most of the day probing the wrackline for amphipods, worms and insects. Competitors for this space will be sent off with a few swift pecks to the back—much worse if they can be pinned down.

"There's a nest and two plover right up there, two o'clock, in front of that clump of high beach grass." Eric stopped the truck and the two researchers ducked behind the nearest dune so they could focus their binoculars on the nest.

The female plover had arrived in late March and found this male strutting like a little Napoleon on his swath of beach. She admired the bold black band across his chest and how he puffed himself up and stomped his tiny feet to get her attention. He had dug several small depressions in the sand and now he was busy tossing stones in the air and flying in tight little figure eights to impress her. Evidently, it was enough to convince her that he would at least be an amusing mate, if not a dutiful father.

As the researchers watched, the female tried out several of the scrapes until she found one that felt comfortable. She surrounded it with shells and bits of debris so the couple could recognize it when they returned from feeding on the featureless shore. Once she selected and decorated the scrape, she turned her attention back to the male who was doing his little macho thing, stomping his feet and goose-stepping back and forth in front of her. Who could resist such blandishments? She must have been attracted because she allowed him to jump on her back and inseminate her with several short, quick ejaculations.

In a few days the researchers returned to find the female sitting on the couple's four new eggs. They set up their banding equipment and waited. When the female flew off to get some food they placed one of Laurie's self-designed traps above the nest. It was ridiculously simple, like something you might see in a Tom and Jerry cartoon, just a wire box propped up over the nest with a stick. They had tied a long string to the stick and ran the string back to their hiding place behind the dunes.

Chapter Seventeen

As soon as the female sat back down on her eggs, Laurie pulled the string and the box fell down over the nest, trapping the female and eggs inside. Laurie held the female while Eric clamped a wire band on one of her legs, then they traded places and Laurie put several color-coded plastic bands on both legs and sealed them on with a glue gun.

Piping plover had just been listed as threatened on the Endangered Species List. They were down to only three hundred birds in Massachusetts, and Eric and Laurie had already banded almost two hundred of them.

The metal band held the official Fisheries and Wildlife identification number so they could keep track of where she had first been tagged. The color-coded bands allowed researchers to report on her progress as she traveled to her wintering grounds. These colored plastic bands also allowed amateur birders to identify individual birds as male or a female. If a spring storm destroyed a female's nest the birders could tell if she had found another male to mate with.

When researchers started studying the migratory patterns of plovers, it looked like most of the plover that nested in New England spent their winters in the Carolinas. Actually, reporting in the Carolinas was strong because there were so many good birders there. It wasn't until in 2001 that researchers discovered twenty five pairs of plovers in the Bahamas. By 2006, they had counted three hundred nineteen pairs in the Bahamas, making this the primary nesting area for New England's piping plovers.

When the plover were put on the endangered species list it meant that there was a new a pot of federal money available to protect them. Coastal communities could suddenly afford to hire monitors to put up symbolic fencing to keep people away from the plover nests, and build exclusionary devices to protect the nesting birds from predators. The exclusionary devices were thirty-foot-diameter cages made out of turkey wire that the monitors built around each nest. They had to build the cages in less than thirty minutes, so the mother could return to her eggs before they got too cold.

Sometimes, people didn't quite get the concept. One monitor found two children sitting inside one of her exclusionary devices. When she asked the mother of the toddlers exactly what they were doing in the cage, the mother said that the concierge at the Chatham Bars Inn had told her there were childcare facilities on the beach.

Some predators learned to associate the exclusionary devices with food. A great horned owl learned that if he landed on top of one of the cages and hooted and hollered for a while, a piping plover would come scooting out the bottom, and the owl would have his dinner. He was like a high school friend of mine who discovered that if you kicked the school's vending machine in exactly the right spot, a coca cola bottle would come trundling out the bottom.

Once the plover were placed on the Endangered Species List, coastal communities were required to restrict ORV's from driving on the beach during the day, when the young, sand-colored birds were still flightless, almost impossible to see, and would often huddle in the ORV tracks to avoid the sun.

The measures worked remarkably well. Within twenty years the numbers of piping plover had increased to several thousand birds. But the biologists were winning the battle, only to lose the war. ORV's started sporting bumper stickers that read, "I love plover, they taste just like chicken."

There were several reports of ORV drivers stomping chicks to death. One driver started to compare piping plover to vermin no better than the virus that causes AIDS. He was a hunter and boasted that he would be happy to be remembered as the person who shot the last female of the endangered species.

The ORV drivers had their own point of view. They were on their summer vacations; they had spent over two hundred dollars to purchase an "oversand permit," but during about four weekends in late June and July they often had to stay in special parking lots at the heads of the beaches and walk to their fishing grounds. This made it difficult for older or disabled fishermen.

The situation had reverted back to the original problem that scientists had debated before the enactment of the Endangered Species Act. Did it make more sense to protect the beach or the birds that lived on the beach? Should you protect the habitat or the species? People who presumably loved beaches and should be their strongest defenders were becoming piping plovers' staunchest enemies. The wound would continue to fester.

The 1987 inlet. For the first time in one hundred forty years the full force of the Atlantic was being hurled at Chatham's mainland.

Chapter Eighteen
The "Old" Inlet
1987

The founder of Eldridge's Tide Charts, George "Chart" Eldridge, called Chatham, "a veritable town at sea, lying further oceanward than any other town in the United States." His native home is also the most erosion prone village on all of Cape Cod. It sits only half a mile behind North Beach, its ephemeral barrier that is overwashed by almost every major storm.

The sound of the sea has always permeated the town. As the dual lights of the lighthouse sweep through the ever-present fog, the surf can be heard from every corner of the village.

But the sound changed after an overnight nor'easter on January second, 1987. Some old timers could hear the difference and woke their wives to make them listen.

The new sound was the more intense roar of the ocean as it tore into the dunes that had just yesterday held the fragile barrier beach together. The storm had flattened the dunes on North Beach, allowing tons of sand and water to rush into Aunt Lydia's Cove.

As the wind died down and the tide turned, the water that had rushed into Pleasant Bay now wanted to find its way back out to sea. The outgoing tide scoured a foot-deep channel through the dunes, down the center of the overwash plain. Each successive outgoing tide then scoured the breach wider. Soon, an eighteen-inch stream meandered back and forth through an overwash plain that was as wide as a football field is long. Its waters were still wadable but they flowed unimpeded into Chatham Harbor. A new inlet had been born.

Chatham was shocked, but it should not have been surprised as scientists had long predicted this scenario. Graham Giese, a coastal geologist at the nearby Woods Hole Oceanographic Institution, had used Demer's 1600's chart of his pre-colonial explorations, William Bradford's 1626 chart of the Sparrowhawk, and Southack's 1717 chart of the wreck of the Whydah to calculate the erosion cycles of the Nauset Beach system.

Using these charts, he showed that for at least the past four hundred years, a breach occurs in the barrier beach off Chatham every one hundred forty years. The south part of the beach then becomes a relic island, cut off from its upstream source of sand, and migrates toward the mainland by rolling over itself. As the relic island moves toward the mainland, the north part of the barrier beach continues to flatten and grow south at the rate of about half a mile a year. After a hundred and forty years it eventually becomes a six-mile-long beach, so low that almost any winter storm can break through it to start the cycle all over again.

But, because Graham was a conservative scientist, he rounded off his figures to one hundred fifty years. Had he stuck to his original estimate he would have only been two days off. If only economists and weathermen could do so well.

But Graham stressed that the most important aspect of the 140 year cycle of inlet formation was that it was being driven by sea level rise. For the past 10,000 years the seas had been rising about one foot every century and causing this 140 year cycle of beach growth and inlet formation. But the big question was, what would happen when the sea started rising to 3 feet every century? Would beach growth and inlet formation happen three times faster or would the barrier beach simply start breaking up?

Chatham fishermen from outside of town were instantly pleased with the new inlet. Now they could cut an hour and a half off the trip to their fishing grounds, saving a few thousand dollars in fuel costs for each trip. But people who lived on the mainland were not so happy. For the first time in one hundred forty years, the village the mooncussers had once called Scrabbletown and now hosted summer cottages was exposed to the full force of the Atlantic.

The summer of 1987 was a time to watch, wait and plan. 1988 would bring the first true test of Chatham's resolve.

On the right, the town of Chatham was given permission to build a seawall to protect its parking lot. In the middle, the Chatham Beach and Tennis Club was given permission to build a seawall to protect its tennis courts. On the left, a private homeowner was determined to be on a coastal dune and was denied permission to build a permanent seawall to protect his multi-million-dollar home.

The Galanti Cottage, January 21, 1988

Chief of Police Barry Eldredge stamped his feet and pulled down his visor against the rain that drizzled out of the low gray clouds. One of his officers beckoned Eldredge to meet him at the crowd barrier. "Chief, this man says he just drove all the way up from Connecticut to take some pictures from the shore."

"Sorry sir, tide's comin' in. I can't let anyone on the beach until it goes back down."

Another helicopter swooped low to film the small crowd gathering on the shore. Journalists in city shoes clambered over the slippery rocks. Eldredge hated having the world snicker at his town and sure as hell wasn't going to be portrayed as some kind of local yokel hick cop on CNN.

The chief knew his town, his tides and his boats. For three

generations his family had made a living from shellfishing and boat building. His great grandfather Oliver Eldredge had started the old Stage Harbor boat yard.

Eldredge didn't like being middled, and it seemed he had been put in the middle of this damn mess ever since August. There hadn't been any major storms, but that hadn't mattered. Moderate winds and the regular high tides had done their damage. By late September they had eroded two hundred feet off the mainland dunes that, after the barrier beach itself, were the second line of defense for Chatham's shorefront homes.

Homeowners were upset with the state, which had recently cobbled together a new wetlands regulation act that included what was known as the dune/bank distinction. This new law said that you could build a seawall to protect your home if you lived on a coastal bank, but not if you lived on a coastal dune.

The state determined the distinction by sending a geologist to scoop up a handful of sand from in front of your home. A look at it under a microscope would determine the size of the sand grains. If they were under a certain size you were considered to be on a coastal bank. If they were over a certain size you were considered to be on a coastal dune. It was a pretty fine distinction if your house was on the line.

The Galanti cottage had been determined to be sitting on a coastal dune, and so it was just a matter of time before it would be washed away.

Eldredge couldn't blame Galanti for being pretty damn angry. In his line of work the chief had seen people lose their homes to flood, fire, wind and storm. It was never easy, and this time it was made a hell of a lot worse by the pack of sensation-seeking journalists vying to get the first shot of one of Chatham's houses falling into the ocean.

The last two weeks had been pure chaos. First, a judge had slapped the state with an injunction that allowed homeowners to truck in boulders; then she had reversed herself. The next day Chatham was full of rumors that the homeowners would try to sneak a convoy of trucks into town anyway. Town counsel had ordered the police to stop the trucks, but somehow they had sneaked in at dawn and dumped the boulders on the beach.

Now the shore was swarming with journalists. Antennae from their broadcast vans swayed above the lighthouse parking lot like masts above a harbor. It was as though a sadistic death watch was taking place.

Eldredge watched as demonstrators passed out literature and painted graffiti about various members of the Conservation Committee on the side of one of the doomed buildings.

An anchorwoman, tottering on her high heels, interviewed a lawyer in front of the Galanti cottage, when suddenly the house shuddered, creaked and slumped into the water.

"Whoops!"

"Damn!"

"Oh, shit!"

The brick chimney clattered to the beach like so many children's blocks. The Atlantic had taken its first victim in over a hundred years. Eight more homes would soon follow.

In the end the dune bank distinction proved to be too arbitrary. One homeowner was determined to be on a coastal bank and was given permission to build a seawall and save his house but his immediate neighbor was determined to be on a coastal dune, therefore condemned to lose his home. Lawyers argued that the law was unconstitutional because it gave the state the power to take a person's home without just compensation as prohibited by the fifth amendment. The case made it all the way up to the Supreme Court before it was turned down.

For it's part, Chatham hired a single person to head up their Coastal Resources Department to try and avoid the lawsuits and confusion that had plagued the town as it tried to come to grips with the reality of sea level rise.

The front of the beach rolled over itself during the "Perfect Storm." Note the camps on the right side of the beach.

Chapter Nineteen
"The Perfect Storm"
October 30, 1991

Donna Edson walked quietly down the twin tire tracks that wound through a low swale of *Phragmites* behind her camp. The tall reeds rustled in the rising wind blowing just above her head. It was the closest thing to a forest that North Beach could offer, and it helped quench her thirst for green grass and trees. She remembered hearing somewhere that sailors craved the sight of green vegetation almost as much as the sight of a woman, and she understood the craving.

Donna loved North Beach. It was her "world apart," part ocean, part desert, part marsh. It stretched south from the parking lot in Orleans to the Chatham inlet that had opened in 1987. Indigenous species of toads, rabbits, grasshoppers and skunk inhabited tiny thickets of vegetation in the swales of the ever-shifting dunes.

By day the hollows in the dunes acted like natural solar collectors. The sun reflected off their sandy parabolic sides, concentrating its rays on the floor of the depression. In the summer, the dune sands reached foot-scorching temperatures rivaling those of many deserts.

The beach grass held this fragile world together. An unseen latticework of underground roots anchored the spines of the shifting dunes. But if disturbed the root system could fail and winter winds would carve blowouts in the dunes. This created a variegated landscape of blowouts, hollows and swales which had been colonized by beds of wild roses, thickets of bayberries, pockets of reeds and the seemingly-fragile vines of morning glories. They were yet another testament to the resiliency of nature amidst a seemingly harsh environment.

This habitat was ideal for numerous species of endangered animals. Piping plover and least terns nested in the new overwash areas, and arctic terns nested on the new sand islands. Equally endangered were the humans who called this stretch of sand and stunted vegetation home. They lived in small clusters of camps that snuggled into the flanks and hollows of the dunes. The Edsons lived in what was unofficially called South Village.

By custom, when North Beachers were in their camps they flew battered flags to signal their presence. On October 30, 1991 only two flags were flying, one over the Edson's camp and one over the camp of their neighbor, Patrick O'Connell.

Donna paused to listen to the long reedy stalks as they jostled in the wind. It was a scratchy, ethereal, frightening sound that matched her edgy mood. It was time to move on, time to leave North Beach. She felt the same nervousness that she imagined the terns felt before they started their long migration southward. The week before they left for good, Donna would watch the entire flock rise up and settle down again at the slightest provocation.

Donna looked forward to the green grass of Florida and the four orange trees that she and Bruce talked about as if they were their children.

Living on North Beach was an acquired lifestyle, but it was one that some Cape Codders had been cultivating for generations. The marriage of the Edsons had united two old Cape Cod families, the Mayos and the Nickersons. Donna had grown up in her father's old camp, "Fort Mayo," and Bruce had inherited forty acres from his grandmother on the Nickerson side.

The days of using the beach marshes for market hunting and cutting hay were long gone. The former hunting camps were now about the only place that locals could go to get away from the tourists and summer people. Most of the North Beachers had a deeply ingrained distaste for crowds and regulations. The irony was not lost on them that ever since the creation of the Cape Cod National Seashore, they were living in one of the most regulated spots on the globe. Many who had once owned their camps now leased them from the Seashore.

The Edsons had been migrating from their camp on North Beach to their mobile home in Florida ever since Bruce had lost his job as a senior pilot with Air New England. The regional airline had crashed financially, causing it to change its local colloquial name, from "Scare New England," to "Kamikaze Airlines." Always the captain, Bruce still rose every morning at five o'clock sharp to catch the tide or just avoid the noonday sun.

Bruce had become the unofficial mayor of North Beach. He kept everyone's camps in repair, made a living shellfishing, and watched over the flats in his role as deputy shellfish warden. Some had suggested that putting Bruce in charge of the clam flats was like putting the fox in charge of the chicken coop. But Bruce had, more or less, risen to the occupation.

Bruce loved the wild beauty of North Beach and only visited the mainland with reluctance. But Donna liked to get off the dunes. She drove into town every day to work in a lawyer's office in

Chatham. She told Bruce it just felt good to put on a pair of high heels every once in awhile.

By October thirtieth, the wind had been blowing steadily for three days. But the Edsons weren't concerned. They had weathered many nor'easters and the radio hadn't reported anything particularly exceptional about this storm. Donna climbed to the top of one of the dunes to join Patrick O'Connell.

Below them, twenty-foot waves assaulted the face of the dune. Even though they were standing a foot apart, they had to yell to each other over the roar of the pounding breakers. As the tide rose, the waves became higher. Finally, one wave split the Holleck's camp in two. Half of the camp stayed in place, while the other half was carried into the middle of the beach. Another wave undermined the Lund camp and it was carried out to sea. Donna couldn't watch as a third camp broke apart and was washed away.

"I have to go back and help Bruce move the cars into the dunes. Why don't you have supper with us? It's too damn scary to be out here alone."

Dinner was short. It was just too exhausting to yell over the roar of the surf. The hurricane lamps flickered and the camp shuddered with the impact of every wave, but the diners were still not concerned. This nor'easter would blow itself out like all the others.

Finally, they abandoned all pretense of talking. Patrick thanked them and promised to light his lantern so they could see he had made it back to his own camp, which was only a hundred feet away. They watched the back of his yellow rain slicker as he trudged through the howling wind and lonely dunes. Within minutes the slicker disappeared in the gray wall of rain and spray torn from the tops of the raging surf. At last Donna saw the tiny flicker of Patrick's lamp. He had made it back. Exhausted, the Edsons fell into bed.

Halloween day broke warm and gray. The wind was still up and ominous tropical warmth now filled the stormy skies. During the night Hurricane Grace had unexpectedly backed into the nor'easter providing it with additional energy. The two storms stayed locked in a terrible embrace, pelting the shores of New England. This was the "perfect storm" that Sebastian would make famous in his book that would also become a movie.

Bruce arose at his usual hour and went outside. Donna was still exhausted from three days of shouting over the sound of the wind and surf. She wanted another hour of sleep before driving down the beach to work.

"Get up, you got to come out and see this."

Donna pulled on her robe and rushed out the door.

"Oh my God!"

All the Edsons could see was water. Most of the dunes had been flattened; most of the camps had disappeared. Water was everywhere. It wasn't just water from a cresting river or water from an offshore storm. It was the full force of the North Atlantic, and there was no guarantee it would subside.

The front of their Jeep Cherokee was in the water and the back of it was buried in the sand. Donna rushed inside to call her daughter on the new cellular phone they used every morning to keep in touch with the mainland.

"We're fine and Patrick's flag is flying, but you wouldn't believe it. The camps are all gone; they all tipped over."

"The Coast guard just called to ask if you're alright. They said they're too busy fixing their boats and dealing with the harbor to get to the beach unless it's an emergency. Do you want to get off"?

"We'll be OK. Bruce'll get the truck started and launch the boat. He ought to be able to make it over to Ryder's Cove."

It took an hour to start up vehicles and get them out of the sand. Bruce hitched both vehicles together and managed to haul the twenty-foot workboat out of the dunes where they had left it for the night. After a few tries, the outboard started and Bruce headed across the bay. Soon, eight-foot waves were slopping over the transom. He had to throw the anchor overboard and swim back to shore.

Dripping wet, Bruce and Donna huddled in the kitchen and watched the boat as it dragged its anchor. "It's gonna sink right in front of our eyes."

Meanwhile, a rescue effort was being launched on the mainland. Skip Stearns had contacted a team of North Beach camp owners. He would lead a caravan of them down the beach at low tide. But halfway out, Deputy Chief of Police Wayne Love told them he would have to turn back, explaining he couldn't risk losing the town truck. "Now, I'm not gonna stop you guys, but fer Christ sakes, be careful." The caravan pressed on.

It was like seeing the cavalry arrive when Donna finally spotted the line of beach buggies threading its way down the ravaged shore. The buggies had to stop in the high dunes and wait while Bruce and Donna waded through the waist-deep water to reach them. All that Donna was able to carry was her pocketbook and their toy poodle, Jingles.

Bruce was nervous the whole way back. Always the captain, he was used to driving himself down the beach. The three buggies had to race the incoming tide. They drove fitfully, stopping and starting between the base of the dunes and the Atlantic Ocean. They would race down the beach as one wave receded, then drive up the face of the dune to wait, while the next wave lapped at their wheels.

By late afternoon, the three beach buggies, with their exhausted occupants, made it to the mainland. Of the seventeen camps in the village only three had survived, the Edsons, Patrick O'Connell's and the Angells. Florida would look especially good this year.

Chapter Twenty
Life's a Beach
North Beach, 2002

When Steve Batty decided to spend the winter in "Second Wind," the North Beach camp he shared with his brothers, he didn't have much choice. He was broke, divorced, homeless and had just suffered a stroke. It was not the life trajectory he had planned after graduating from Brown and then Harvard Business School. But the beach was not a bad place to catch your breath. The rent was free, he had all the shellfish he could eat, and the town gave him a small stipend to watch over the clam flats as the deputy shellfish warden.

Steve soon discovered that there were three well-defined groups on the beach. There were the camp owners, many of whom had owned their camps for several generations. There were the people who came out almost every weekend and stayed in their self-contained vehicles. Many of these folks were dedicated hunters and fishermen whose families all knew each other and had also been coming out for years.

The third group drove out for the day. These were the weekend warriors, the "lazidiots," who drove too fast, got stuck in the sand and didn't understand the etiquette of the beach. When the trails were flooded they drove up through the dunes, endangering the shacks. They used camp owners' private outhouses or simply did their business in the dunes. They had also been known to break into camps and steal personal mementos and whatever liquor they could find. The number of camp owners and self contained vehicle operators had been steadily declining, while the number of "lazidiots" had been steadily increasing.

The reasons for the shift were many. It took a lot of time and know-how to keep a camp running. You had to maintain a small flotilla of vehicles and boats to get on and off the beach. Repairs were a constant problem, the owners were getting older, and their children were leaving for other parts of the country.

Physical prowess was held in high esteem on the beach. Mac MacCausland had Scottish Highlander blood from one side of his family and blood of one of the warriors who had fought in King Philip's War on the other. Though he was getting on, every fall Mac would drive south with his crossbow and bag five or six deer from every state where it was legal.

Relations were generally good between the first two groups. Much of the North Beach social life revolved around a camp owner called Baldy. Baldy always had a beer in his hand. If he liked you, he and his wife could be hugely generous; if he didn't, watch out.

Kids remembered Baldy because he would do things like cut a live sand worm in two and put one half in each nostril so it looked like the worm was crawling in and out of his nose. Mac remembered Baldy for his Herculean strength. Even as an older man Baldy he had deeply angular muscles that were harder than the wooden deck on his camp. Mac figured he must have carried his excavators from job to job instead of trailering them in a ten-wheel dump truck.

One day a group of weekend warriors sped past Baldy's camp on Middle Road. Mac remembered, "He intercepted them when they were leaving to instruct them on the speed limit. The truck was filled with six or seven, twenty five-to thirty-year-old drunken lads. One rugged fellow was standing in the bed of the truck giving Baldy some lip. When he called him an old man, Baldy went vertical like a Polaris missile and at the apex of his jump slapped the guy on the side of his nose. His buddies said, 'Go down; you can take him.' Baldy stood there with his level palms up, a sort of Yule Brenner squint in his eyes and his head tilted to the side. As the sting became numbness the lad quietly said, "I don't think so." They drove away at an idle, well under fifteen miles per hour."

But Steve found there were also some other divisions on the beach. After what Cape Codders still prefer to call the "No-Name Storm" instead of the more pumped up "Perfect Storm," Steve's old nemesis from his days at Northfield Mount Herman Academy moved his camp from South Village to North Village and started to throw his weight around. Bill Hammatt was a local lawyer, so the town had appointed him head of the North Beach Advisory Committee.

Hammatt also set up a legal trust to limit liability on his camp, and eventually brought in Scott Morris as a partner. Scott was the long-time head of the Massachusetts Beach Buggy Association's access committee and would eventually become its president. The MBBA had changed since it had started in 1950 from being a quasi-environmental group that supported the creation of the Cape Cod National Seashore and taught its members to respect the beach to an organization that was primarily interested in maintaining access to the beach in the face of tighter environmental regulations. Its bête noir had become the piping plover, which had nested unmolested on these beaches for tens of thousands of years.

The No-Name Storm had also left the beach low and prone to overwashes. These overwashes often destroyed the beach buggy trail that wound down the middle of the beach so drivers had to drive their ORV's on the beach between the dunes and the ocean. The camp owners who lived in the south of North Village wanted to keep it that way, but the owners of the camps in the north of the village didn't want beach buggies driving in front of their camps. So Bill Hammatt quietly arranged for the town to rebuild Middle Road through the dunes.

But the trouble with the new configuration of Middle Road was that it went straight through the dunes and the properties of the Battys, the Trueloves, the Harrisses and the Broads. This made their camps more prone to erosion and plagued by the hordes of lazidiots. Nobody wanted their grandchild to be hit by an inebriated driver while playing in the sand in front of their family camp.

The dispute became something like the Hatfield's and the McCoy's, without the charm. There were allegations that some people had buried old vehicles in the road to puncture tires, and threats that people would be burned out of their homes. Sand had been poured into the locks of one owner's camp. Steve supported his friend Russell Broad, accompanying him to town hall on several occasions.

It was understandable that emotions were starting to run high. The beach was coming to the peak of a thirty-one-year cycle of astronomically high tides. This was when most erosion events happened. It was the time when you saw the most articles about sea level rise. The two forces would soon come together to open a new chapter in the history of Chatham's North Beach.

It was a history that Steve Batty would mostly experience from afar. In 2005 he left the beach, relaxed and renewed, "I don't think there is any place else on earth where I could have experienced such God given mental and physical health benefits."

The old inlet was constricting water flowing out of Pleasant Bay. This led to the creation of the new inlet opposite Minister's Point on the right hand side of this photo.

Chapter Twenty One
The New Inlet
April 24, 2007

Graham Giese and Paul Fulcher drove down Nauset Beach to see the damage from the Patriot's Day storm. The two made an interesting couple. Graham was the reigning geological expert who had used the rare historical nautical charts to figure out the one hundred-forty-year Nauset Beach cycle.

This was the first time he had been able to get out to the beach since the storm, but he had already been quoted in the papers as saying he felt the break in the shoreline would probably close.

Paul was the Orleans superintendent of beaches who had grown up surfing on Nauset Beach. Even as a kid, he was always the first to figure out where the surf would break. He seemed to know in his bones the ephemeral patterns this flux of energy would form. Offshore sand bars, coarse-grained sand, even an imperceptible bulge in the beach could cause the surf to break in a new location.

Perhaps it was his intuitive appreciation of tidal energy that made Paul dubious of the more episodically-inclined expert's prediction. But he had also been observing the new break and overwash area every day since the storm. Of course, it was part of his job, but he also did it out of pure curiosity.

He had seen that the day after the April fifteenth storm, the water that flowed through the new overwash area had seeped back into the beach at low tide. The next day, it was scouring a tiny little channel of water that continued to trickle back into the ocean, even at low tide. In another day, the channel was three feet deep and the overwash fifty feet wide. If this was a new inlet it would directly affect the houses on North beach and the mainland.

A week after the storm, Graham called Paul to ask for a ride to see the overwash area. They met at the Nauset Beach parking lot in the early morning and proceeded down the still ravaged beach. Paul liked the genial scientist and ribbed him gently about getting so much ink in the local papers.

"So where do you think this storm will go down in history, Graham?"

That was a damn good question. The recent Patriot's Day Storm certainly hadn't been as powerful as the No-Name storm or any of the hurricanes in the Fifties, and it had packed far less punch than either the 1958 or the 1978 nor'easters.

However, sea level had risen about three inches every twenty five years, so the storm had occurred when the sea level was half a foot higher than it was during those former storms. Plus, it had occurred right after winter storms had lowered the barrier beach in several vulnerable points. What's more, they were approaching the peak of the Proxigean tides when the moon was closest to the earth. To make matters worse, the Patriot's Day Storm had hung around through several cycles of the highest spring tides. All of these factors together had caused what Graham called a "super overwash situation." The big question on everyone's mind was whether the shallow channel that was now running through the overwash area would close back up or become a new inlet. People were already talking about spending millions of dollars to fill in the break so that wouldn't happen.

Paul suspected that Graham already knew it was too late. The pattern of coastal organization had shifted. Writers could write, experts could opine and politicians could push to fill in the channel, but nature was making it clear that a new era in the beach cycle had begun.

Paul could sense the scientist's discomfort. He had already been quoted in the papers as saying that the channel would fill back in on its own. How could he back down now?

"So Graham, how about a little wager?"

Graham laughed.

"Say about a hundred bucks that the inlet will stay open?"

Graham winced.

"How about a cup of hot, black coffee instead?"

Graham was silent on the drive back . Why had he been so damn positive in his earlier quotes? He had just seen how well established the channel was, and that it wouldn't take much for it to become the new inlet. He would have to make a hundred-eighty-degree turn in full view of the public. But he knew that part of being a scientist was to admit you were wrong in the face of better evidence.

Paul pushed the issue a little further, "So what do you think about this idea of filling in the break?"

"Oh God, I don't know how I'm going to handle that one."

That summer, the voters of Chatham would wisely reject a four million, two hundred thousand dollar plan to fill the inlet, and many would enjoy the newly-opened area. It became a favorite place to land a boat and spend the day fishing and swimming and naturalists took tourists to the site to see gray seals that had started using the inlet to get in and out of Pleasant Bay.

Adult gray seals can weigh up to eight hundred pounds.

The Attack

It was the end of another long hot summer day. A lone gray seal surfed the five-foot waves rolling through the new inlet and across the broad plateau of sand stretching half a mile into Pleasant Bay. The currents carried the seal swiftly over a dizzying tableau of sand, crabs and scattering minnows. He cruised along, gulping great mouthfuls of sand eels.

The sand eels were one of the first species to discover the new inlet, with its concentrated supply of plankton. The gray seal,

Halichoerus gryphus or "the hook-nosed sea pig," had followed them. The entire herd of several hundred seals had moved north from their old feeding grounds off Monomoy to feed on the shoals of sand eels sluicing in and out of the new inlet. The male seal preferred to feed alone in the inlet, but now it was time to rejoin the rest of the herd sitting safely on Lighthouse Beach.

The tide had started to slacken, and rays of the setting sun were now entering the water at an oblique angle. The changing tide slowed the seal down and the setting sun impeded his vision, which was specially adapted for daylight hunting. He had already eaten over seventy pounds of bass and eels. His body glistened with more than enough fat and blubber to carry him through the upcoming breeding season when he would be too busy coupling and competing with other males to gorge on sand eels.

The seal swam slowly out of the inlet and into the inky green waters of the Atlantic. Here he turned south to swim parallel to the shore to join the rest of the herd now sunning themselves quietly in the fading light on Lighthouse Beach. But something was wrong. The water was deathly quiet. He no longer heard the clicks of small fish feeding on the bottom. He looked into the gloom, but his daylight vision failed him. The advantage had shifted to his enemy lurking almost motionless and invisible against the bottom.

The great white shark had lain in ambush just beyond the inlet for half an hour. She had heard the muffled laughter of human surfers overhead, but they were not her favored prey. Now, as the long dark silhouette of the seal appeared above her, the shark was ready. Muscles along the caudal peduncle of her tail powered her skyward. Nictitating membranes slid in place to protect her eyes during the attack. She rolled back, opened her mouth, and drove the full force of five thousand pounds of rasping skin, cartilaginous flesh and a mouthful of serrated daggers into the seal's vulnerable underbelly. The impact of the blow threw the seal into the air and he fell back into a cloud of his own frothing red blood. The shark swam back toward the bottom and waited. She had severed several arteries in the seal's belly. Better to wait for the seal to weaken before attacking again. The last thing she needed was to have an eye gouged out in the death throes of a large, potentially dangerous prey animal that was going to die anyway. The shark could smell the cloud of blood billowing toward her and feel vibrations of the seal's still-vigorously-pumping heart. But she knew its struggles would soon be over.

When it was clear the seal was too weak to fight back, the shark circled in for the kill. She tore three great mouthfuls of viscera and blubber out of the still twitching animal, then abandoned the carcass. She was not really all that hungry. She had just enjoyed the hunt off her newly-discovered inlet. Onshore, a group of surfers looked on in horror. They had been in that same water only moments before. The partially-eaten carcass would wash ashore the following morning.

After feeding, the shark dove back to the bottom and continued to swim slowly south. Was this the same young female shark that a boatload of tourists had seen thrashing around in a Monomoy tidepool three years before? A few weeks after that she had blundered into a tidal creek on the Elizabeth Islands, thirty miles away, a creek where well-tanned summer kids used to tie ropes to the Naushon footbridge and water ski on the powerful currents of the outgoing tide.

Evidently, this accident-prone young shark just loved shallow water, fast currents and, perhaps, people. She received more than her dose of all three as she stayed stuck in the shallow Naushon tide pool for two weeks while photographers paid entrepreneurial Woods Hole fishermen a hundred bucks an hour to ferry them across Vineyard Sound to take pictures. State and federal officials would only allow photographers to get within a hundred feet of the shark as she quietly cruised about her little pond. Ignoring this rule, one scientist noticed that she liked to surface in one spot when she swam around the tidal pool. He stationed himself over the spot and patted her on every circuit.

Normally, only humans get such paparazzi treatment. But this young shark had picked up this curious new behavior in order to find and eat the abundant and easy-to-catch food near shore. Such roguish behavior had its benefits, as long as you didn't mind spending a few weeks in a pond having hundreds of people take your picture.

This was not the first time a shark had developed such behavior. During the 1930's a bull shark wandered in and out of shallow creeks along the Jersey Shore. There, it attacked mostly young children swimming in the warm protected waters. For several weeks the entire East Coast was transfixed as the shark killed swimmer after swimmer during a summer of terror. Sound familiar? Peter Benchley used the incident as the basis for his best seller, *Jaws*.

But was this really the same shark that had blundered into a Monomoy tidepool three years before, then been seen attacking seals off Monomoy last summer, then Nauset and Lighthouse beaches this summer? Did we encourage her to come back by allowing her to become habituated to humans when she was trapped on Naushon? We will never know. Just before scientists coaxed the shark out of her brief captivity on Naushon Creek they jabbed a tag into her tail. The last thing they saw as the shark swam back into Vineyard Sound was the tag caught in a bed of eelgrass.

The Harris camp was lifted off its foundation and moved thirty feet by the storm surge created by Hurricane Noel.

Chapter Twenty Two
Hurricane Noel
November 3, 2007

When you think you have finally figured out one of nature's patterns, she has a way of throwing in a twist just to keep you humble. After finishing a chapter on the formation of the 2007 inlet, I paused to read the morning papers. It was seven months later and Hurricane Noel was moving north from Florida. It had killed one hundred forty eight people in the Caribbean and then been downgraded to an extratropical cyclone. But it was still packing plenty of punch and accelerating rapidly north.

I was working on a film about the inlet and planned to meet the camera crew half way to the Cape. I packed my bags, rented a car, then realized it would be foolhardy to drive to Cape Cod in hurricane-force winds. Besides, I could probably learn just as much watching television in my warm house in Ipswich as I could driving into the teeth of the oncoming storm.

I set up a "situation room" in my attic office. One computer carried emails from crew members who were stuck filming the storm in Scituate and Cohasset; another computer carried images from a camera on the mainland in Chatham that was trained on the camps on the outer beach. My television was tuned to the Weather Channel's Mike Bettes, reporting from the Chatham Fish Pier.

I watched the Weather Channel all day. It was excellent at providing the big picture. They had crews stationed on the beaches of Florida, North Carolina, New Jersey, Montauk Point, New York and Chatham, where fishermen were hauling in huge striped bass, made hyperactive by all the oxygen-saturated waves that were churned up by the storm.

But the Weather Channel was frustratingly lax at presenting details. Here they were, stationed in Chatham, and they seemed totally nonplussed about the most visually striking impacts of the storm.

While their state-of-the-art cameras showed fairly mundane scenes of moored fishing boats bobbing around in slightly choppy seas, our own little "inletcam" was transmitting far more compelling images of camps on North Beach surrounded by ever-rising waves. Although the shots were rain swept and blurry, you could see Donald Harriss' and Russell Broad's camps still standing, but perilously close to disaster. I even tried to call the Weather Channel headquarters to tell them what was happening only half a mile from their cameras, but there was no way to reach an actual human being in Atlanta on a Saturday morning. I wonder how many tips the Weather Channel misses by not having someone available to take calls during off-hour emergencies.

On Sunday morning I awoke at four a.m. and drove to Plymouth to meet the rest of the camera crew and decide how we would proceed down to the Cape. Susan Haney had spent all day Saturday filming Noel in Scituate, and Bob Daniels was recovering from a bumpy overnight flight back from D.C. We convened over coffee at Dunkin' Donuts and decided to drive to the Chatham airport to see if we could hire a small commercial sightseeing aircraft to fly us over the beach. If that didn't work, we would drive on to the fish pier to see if we couldn't find a boat. If all else failed, we could always drive to Orleans and hitch a ride down the beach in a four-wheel-drive vehicle.

I started to get an inkling of what our day would be like at the airport. The flight operator had put his planes into a hangar the night before, and now the power was out, so he couldn't open his hangar door.

"With fifty thousand people out of electricity, I don't think we will be back in business anytime soon."

We flagged down a Chatham cop, who scrolled through his cruiser's computer readout and told us no beach patrols were scheduled for the entire day. A news crew at the lighthouse overlook had already given up on trying to get to the outer beach. I was beginning to think I had led everyone on a wild goose chase; the entire day was going to be a bust, and an expensive one at that.

But our luck finally started to improve. There were camera crews from all the Boston stations on scene to film the results of the storm. One of them was filming the top of the Broad's camp, half a mile across the harbor. But Russell Broad's roof was not on his house. It had been washed a mile down the beach and was now lodged beside a grounded fishing boat. A recreational boat had also been swept through the inlet and into the open Atlantic, but then had been swept back in twenty-foot seas and eighty-five-mile-an-hour gusts, to come to rest on the foot of South Beach. Noel had come much closer to Cape Cod than anyone had expected.

I flagged down the assistant harbormaster's boat. He agreed to take us to the outer beach to snap a few pictures. In these kinds of emergency situations everyone pitches in to help. But he cautioned, "It has to be a quick tour. The tide is falling, and there is so much sand in the water you can't see if there are any new shoals."

It was a small step in the right direction. We leapt at the chance.

As we approached the shore I could see that the Broad's house was scattered in several pieces and the Harriss camp was still upright. But my attention was really elsewhere. There were several people scouring the beach. I yelled across the water to see if anyone could give us a ride back down the beach. Nobody could hear. There was only one solution. I apologized to Susan, stripped to my skivvies and waded ashore.

On shore, John Kelly was amused at my audacity when I asked him for a ride.

"Sure. We have a few more things to do in our camp, but if you promise to put your clothes back on, I'll drive you to Orleans in my pickup."

Under the circumstances, it was not a bad deal at all.

The harbormaster dropped us off and I finally had a chance to inspect things more carefully. It was totally disorienting. I couldn't tell where Russell's two-story camp and boathouse had stood. The foundation was almost in the water, the kitchen was lying on the beach, the boat house appeared to have been swept toward the ocean while the main house had crashed into Russell's bulkhead, broken apart, then been swept into the bay, where it lost its roof.

Donald's house was still standing, but canted over toward the ocean. The kitchen had torn off the old part of the camp and stood upright in the sand. The tattered flags that all the camp owners used to announce their presence still fluttered from the ceiling over the cooking utensils, which still hung from the wall as if nothing had happened. Beside them was the refrigerator, still upright and attached to the wall, holding shelves stacked neatly with clean plates.

It was the refrigerator that made me into one of the most loathsome creatures on the earth, a looter. My back was still aching from the morning drive and I had an aspirin in my camera bag, but no bottled water. There was the refrigerator. There was my pill. I opened the door, and sure enough, there were rows of unopened ginger ale cans. It didn't even fizz when I flipped open the can. I swallowed my pill and turned to offer a quiet toast to Seamore Nickerson and the four generations of Harrisses who had all loved this camp.

The rest was not a pretty sight. All the chairs and rugs that Donald had stacked on tables the week before were scattered through the shell of the ravaged camp. Most of the central chimney had tumbled down and lay in big blocks of bricks on the sand-covered floor. Gas lanterns still hung from the walls and a forlorn little sign hung outside marked this as The Seamore Camp, circa 1896.

Don's well head and hand pump were still in place but now located on the lower beach instead of in the dunes. The ocean had eaten away seventy feet of dunes in front of Fred Truelove's camp, eighty feet in front of Donald's camp and ninety feet from in front of the Broad camp. All that sand had actually narrowed the inlet and deposited more than ten feet of material in front of the remains of Russell's bulkhead. The current was slack, a teapot and a toilet seat lay, side by side, in the quiet waters of the bay.

The Broad and Harris camps were gone, and the Patriot's Day storm and Hurricane Noel seemed to be the culprits. But really, these storms only sped up the erosion process that would have destroyed the camps in a few months anyway.

And while it was clear that the steady pattern of sea level rise is the engine that causes the inlet to form every hundred forty years, it was getting increasingly difficult to avoid the question of what happens when the pace of sea-level rise quickens.

Chapter Twenty Two

Twenty four feet of sand were washed away from underneath Fred Truelove's camp on a calm December day during a single high tide.

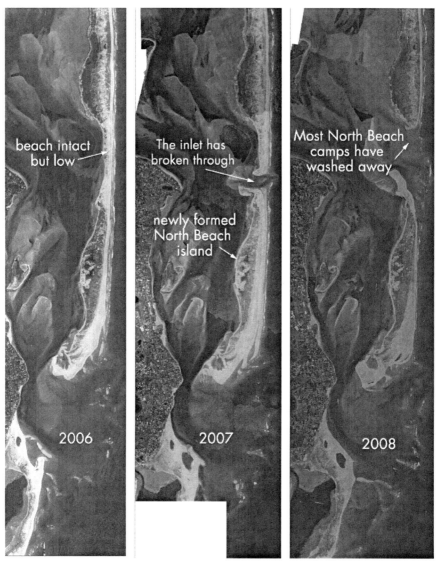

This series of aerial photos show the recent evolution of Chatham's most recent inlet. In 2006 North Beach was still intact but the beach was low. In 2007 the inlet broke through opposite Minister's Point, creating North Beach Island in the center of these photos. By 2008 most of the camps on North Beach had washed away as the inlet migrated half a mile north.
Photo, courtesy of Ted Keon, Chatham Department of Coastal Resources.

Chapter Twenty Three

The remains of the Kelley camp.

Chapter Twenty Three
The Last Summer
North Village, 2009

June 18 was chilly and overcast. Low grey clouds threatened rain, but none ever materialized. A lobsterman from the Chatham Bars Inn offered to drop me on the outer beach while he tended to his nearby pots.

We landed on a massive new lobe of sand that jutted off North Beach into Pleasant Bay. I could hardly believe my eyes. The new sandspit had started growing only seven months before, and now it was half a mile long and consisted of twelve acres of new land. Every day another half acre of subtidal sand would sweep around the tip of North Beach to attach to this spit, which eventually was called Scatteree Island.

Three years ago there were fourteen camps sitting on North Beach like seagulls in a row. But month by month the camps were washing away as the new inlet migrated north at the rate of ten feet a day. The five camps that remained had been moved several times in the past year. After the experiences with the 1987 inlet the town and state were leaning over backwards to accommodate the camp owners. Now their camps sat clustered together on the remaining two acres of Bill Hammatt's rapidly-eroding piece of private property.

Watching the interaction between camp owners and the quickly-changing landscape was like watching the final moves of a lopsided chess match, where one player was cornered by his opponent's most powerful pieces. All he could do was slide his last pawn ineffectually back and forth between the last two red squares in a vain attempt to forestall the inevitable.

I walked out to the last camp at the end of the beach, where the ocean was now rushing past like the Mississippi River at full flood. The camp had been moved here a week ago, but waves had torn five feet off the low dune in front of the camp while another overwash had hemmed it in the back. We were heading into a ten-day period of very high tides, and it would only be a matter of days before all the camps were swept away.

The Hammattyville Gale

On June twenty second, an unseasonable nor'easter struck and lingered through four tidal cycles. It was not a particularly powerful storm, only packing twenty-mile-an-hour winds. If the storm had struck when the tides were still low, it would have caused little damage. However, it hit during some of the highest tides of the year, when the beach was still recovering from the past winter's storms.

The day after the mini nor'easter, Bill Hammatt stood on the storm-ravaged beach. A year ago he had invited the four remaining camp owners to move their camps to what was left of his five-and-a-half acres on the northernmost section of North Village. A local wag had dubbed the new village Hammatteyville.

Now Bill Hammatt's neighbors' camps sat, tilted at odd angles, on the last sandy two acres of his property. Two of the camps had already slumped off the dunes and were awash in the seething waters of the Atlantic. It looked like a child had thrown a temper tantrum and strewn his toys all over the beach. *The Boston Globe* had duly recorded the damage in a dramatic, front page, above-the-fold, news photograph.

Perhaps Bill Hammatt wanted to believe this was just an act of God, the result of a single violent storm. However, the storm had been relatively mild. The ugly truth was that the ocean had been advancing against his land at the rate five feet a day, and would have overcome his property even without the storm. But Bill Hammatt was a lawyer, and lawyers are constitutionally leery of numbers. They would much rather talk in forceful language about vague generalities than to have to deal with real numbers. Numbers don't give lawyers much to argue about.

Three of Bill Hammatt's neighbors— Tod Thayer, John Shea, and John Kelley— all decided that they had enjoyed their camps for several generations and now was the time to let go. They had each spent close to forty thousand dollars moving their camps, the last time only a week before the camps were destroyed. They sadly watched as, one by one, their camps were demolished and put into containers to be barged away.

But, the two remaining camp owners, Copey Coppedge and Dr. Colin Fuller, had much deeper pockets and an unwillingness to let go. They had already spent close to a hundred thousand dollars to move their jointly-owned camp four times in the past year. This time they hired an expensive barge and a ninety thousand dollar crane, which almost had to be abandoned after getting stuck in the storm-tossed beach sand. The two had decided to move their camp, Diastole, to the mainland then to North Beach Island on the other side of the inlet. They would move it back to North Beach after it started growing again and after lawyers determined who owned the beach as the last man standing.

The problem with their scheme was that this was simply not the way barrier beach systems worked. The new inlet had made the southern portion of North Beach into a relic island, cut off from its upstream source of sand. A barrier beach needs this source to replace sand that is continually being washed away by storms and the currents that course along its shore. All that upstream sand now flowed through the inlet into Pleasant Bay, leaving the front end of North Beach Island eroding at the rate of eighty feet a year. The sand was being swept around and over the island, making it ever more prone to erosion. This was the process of rollover as first described by Steve Leatherman in his thesis on Assateague Island which accurately predicted that North Beach Island would roll over and break up in less than twenty years.

It had cost the two families close to a quarter of a million dollars in engineering and legal fees to move their two hundred thousand dollar camp this far; it would cost them at least another fifty thousand dollars to gain just a few more years on the disappearing beach; that is, if they could get permission to barge the camp back to North Beach Island. One neighbor said it looked like they had gone through a nasty divorce and were willing to spend any amount of money to get even. The problem with this divorce was that it was with Mother Nature, and Mother Nature always wins.

Bill Hammatt, who owned the sole remaining camp on North Beach, had thought of demolishing it when his neighbors had done so but had changed his mind and decided to move it one more time, then to just leave it where it was on the beach and hope for the best. His neighbors cheered him on. Of course, doing nothing meant there was always the chance that the tax-payer-subsidized Federal Emergency Management Administration would reimburse him for his loss if his camp was severely damaged by a storm.

On August eighth Bill Hammatt invited a hundred fifty guests to his camp for one final party. Though the party was a roaring success, Hammatt had raised the town's ire once again, because all one hundred fifty people had driven over the dunes, effectively creating a new road through town-owned conservation lands.

But nature always has the last laugh.

Hubris, October 18, 2009

On Sunday October 18th, townspeople gathered on Scatteree Landing to peer through the heavy fog and rain. Half a mile away, Hammatt's Hangar suddenly slumped into the Atlantic ocean. It floated and bumped two hundred feet south down the beach, was swept into the inlet and then bumped and floated two hundred feet back up the bay on the inside of North Beach.

When the tide turned, Hammatt's Hangar was stranded in rapidly-accumulating sand in the center of the inlet. It would remain there for over a week with five feet of water rushing through it at every high tide, and camera crews clamoring over its skeletal remains at every low tide. FEMA eventually paid for the removal of the remains of the last camp on North Beach. It was a difficult process that took several days because the camp was mired in a crater of soft sand and rushing water and the cleanup crew could only work at dead-low tide. It would have been far cheaper if FEMA could have reimbursed Bill Hammatt to remove his camp before it was destroyed by the storm.

There were still a dozen more camps on Nauset Beach in Orleans and eleven more to the south on North Beach Island. But, so many local, state and federal regulations had been bent, broken and ignored that it seemed unlikely that officials would be so lenient about moving camps in the future. As the ocean moved toward North Beach Island it was clear that we would soon find out.

DATE		HIGH			LOW				☀		☽
		AM	ft	PM	ft	AM	ft	PM	ft	RISE	SET	MOON
1	Sun	8:06	4.4	8:46	4.1	2:42	0.6	3:20	0.4	6:18	7:07	
2	Mon	9:04	4.6	9:40	4.4	3:40	0.5	4:15	0.2	6:17	7:08	
3	Tue	10:00	4.8	10:31	4.7	4:36	0.3	5:07	0.1	6:15	7:09	
4	Wed	10:54	5.0	11:20	5.0	5:31	0.0	5:58	-0.1	6:13	7:10	
5	Thu	11:4	5.2			6:23	-0.2	6:46	-0.2	6:12	7:11	
6	Fri	12:08	5.4	12:35	5.3	7:13	-0.5	7:34	-0.4	6:10	7:13	
7	Sat	12:56	5.6	1:21	5.4	8:03	-0.6	8:22	-0.4	6:08	7:14	
8	Sun	1:44	5.8	2:17	5.4	8:54	-0.7	9:11	-0.4	6:07	7:15	
9	Mon	2:33	5.9	3:09	5.3	9:45	-0.7	10:02	-0.3	6:05	7:16	
10	Tue	3:25	5.8	4:04	5.1	10:38	-0.6	10:54	-0.2	6:04	7:17	
11	Wed	4:1	5.6	5:01	4.8	11:33	-0.4	11:50	0.0	6:02	7:18	
12	Thu	5:17	5.3	6:01	4.7			12:31	-0.2	6:00	7:19	
13	Fri	6:19	5.0	7:05	4.5	12:50	0.2	1:32	0.0	5:59	7:20	
14	Sat	7:24	4.8	8:11	4.4	1:53	0.3	2:36	0.1	5:57	7:21	
15	Sun	8:31	4.7	9:13	4.5	2:59	0.4	3:39	0.2	5:56	7:22	
16	Mon	9:36	4.6	10:10	4.6	4:04	0.4	4:38	0.2	5:54	7:24	
17	Tue	10:33	4.6	11:00	4.7	5:04	0.3	5:30	0.2	5:53	7:25	
18	Wed	11:24	4.6	11:43	4.8	5:56	0.2	6:16	0.2	5:51	7:26	

High tide chart for Aunt Lydia's cove with highest tides circled.

Chapter Twenty Four
How to Predict Erosion

When the inlet was formed in 2007, the dozen families at risk of losing their homes wanted to know how quickly erosion was going to affect their lives. Unfortunately, scientists had very little to offer. Sure, they could predict that Cape Cod was going to wash away in about five thousand years, that the Cape might lose a certain number of acres in two hundred years, or what a hundred-year storm might look like. But they couldn't tell a camp owner if he could stay in his barrier beach camp for another season, or tell a homeowner whether it made sense to build a seawall to protect his oceanfront house for another twenty years.

That is because erosion doesn't proceed in a nice linear fashion according to sea level rise. Instead, you have stretches of about seven years when erosion is severe, followed by stretches of about seven years when erosion is mild. From about 2004 to 2011 we saw houses being swept off beaches or left teetering on the edge of cliffs in Chatham, Truro, Wellfleet, Dennis, Nantucket and Martha's Vineyard.

Then, in November of 2011, it seemed as if the erosion stopped. There was no snow, few storms and little erosion. Snowy owls started appearing in all of the lower forty eight states. It was as if ten thousand Hedwigs had entered the world of the muggles, though scientists suggested it had more to do with the billions of lemmings that had been born the year before.

What really happened is that the North American climate had slipped into an intense La Niña period. La Niñas occur when cool waters spread across the Pacific Ocean, changing the way weather systems travel across the American continent. During regular years and El Niño years, cold air bulges out of Canada causing winter storms to swoop south and gather up humid air from the Gulf of Mexico. Then the storms careen up the East Coast, glance off Cape Hatteras and slam into Cape Cod as raging nor'easters that hang around through several erosion-causing cycles of high tides. During La Niña years the weather is mild and any storms that occur march straight across the continent and don't pick up warm, energy-producing air from the Gulf of Mexico.

But the tides are the main factor that cause the seven-year stretches of erosion. When the inlet broke through in 2007, we were close to the peak of the proxigean tides. The word proxigean comes from "proxi," which means close or proximate and "perigee" which means closest to the earth and lined up in a row. Put the two words together and you have proxigean, worth a good thirty points in your next game of scrabble.

Extreme proxigean spring tides occur when the moon is closest to the earth, the earth is closest to the sun, and they are all lined up as they are during a new or full moon. It is a confluence of tidal influences that occurs only every thirty one years.

For background, there are several factors that influence tides. Spring tides occur every fourteen-and-a-half days, when the moon is new or full. Perigean tides occur about every twenty-eight days when the earth, moon and sun are all in line and the moon is closest to the earth. Then you have proxigean spring tides about every fourteen-and-a-half months when the earth, the moon, and the sun are all lined up and the earth is closest to the sun. These are the important ones that flow through the tide tables like a pig in a python. Finally, you have extreme proxigean spring tides every thirty-one years, when all the elements are combined.

We had an extreme proxigean spring high tide on December 12, 2008. You might remember noticing that the moon looked unnaturally large that year. It was not an optical illusion; the moon was about thirty thousand miles closer to the earth than normal. This helped create severe erosion for about three-and-a-half years before that peak of the extreme proxigean tides, followed by another three-and-a-half years of severe erosion after the peak.

In 1978, a man called Fergus Wood ran through three hundred years of erosion events and found the same seven-year stretches of erosion that occurred during the peaks of the proxigean tides. He wrote a book called *Tidal Dynamics*, but it was largely ignored because Fergus had left the impression that proxigean tides were the primary cause of the events. Today we know that proxigean tides do not *cause* erosion, but they can be the best *predictors* of erosion events that are connected to storms.

So, a storm that occurs during a full-moon proxigean tide will cause a great deal of damage, but if that same storm occurs only seven days later, during a non-proxigean half moon, it will cause considerably less erosion because the high tide might be a foot or more lower.

Of course you can't predict storms more than about two weeks in advance, but you can very accurately predict what the tides will be hundreds of years in the future. This means that a homeowner can use a simple tide table to predict the days, weeks and years when their house will be in danger from erosion.

To predict erosion events, you can simply purchase a tide table at the beginning of every year (or go online) and put boxes around all the days that the tides will go over the second highest point on the table, which corresponds to the height of a typical flood tide. If you are using a Boston-based table you will put a box around all the tides over eleven feet. If you are using a tide table for the Chatham inlet/Aunt Lydia's Cove area you will put a box around tides over five feet. Those boxes, along with a simple weather forecast, will tell you when your house will be vulnerable to severe erosion during a storm.

During non-proxigean years you might have only five days out of a single month when the tides go over that maximum height, but in a proxigean year you might have as many as twenty days when the tides go above it in the same month. This means that you have four times as many days when a storm can cause extreme erosion.

It is always possible that a single storm can occur during a seven-year stretch of lower high tides and still cause considerable damage. This happened during the Perfect Storm in 1991. But what we are interested in here is probability. You are about four times more likely to have significant erosion during a proxigean year when even a run-of-the-mill storm can ride in on tides that are several feet higher than normal. There is another intriguing possibility. Some researchers think that the reason proxigean tides and La Niñas track so closely is that the proxigean tides actually cause the upwelling of cold water in the Pacific.

All of this might sound esoteric, but it's crucial information for coastal homeowner. It can help you decide whether you are likely to be safe for another seven, fourteen or thirty years. The knowledge of proxigean tides can help town officials decide if or when to give homeowners permission to build anti-erosion devices like seawalls or gabions to slow down the effects of sea level rise. And it can also help marine contracting companies know which years may bring brisk business and which years may be lean.

If we use proxigean tides to peer into the future we can see some heartening, albeit short-term good news in that we have just entered a stretch of mild erosion that will last about fourteen years. But it will be but a temporary reprieve. The tides will rise again.

The Audubon Camp, formally known as the Crowell Camp. Notice the set of stairs to the left, which had been torn off the building by waves. This photo showed that the ocean was starting to destroy the small camp and that others would soon be in imminent danger.

Chapter Twenty Five
The Beginning of the End
North Beach Island — August 1, 2011

Ted Keon took Mark Adams across Chatham Harbor by boat to measure the rate of erosion on North Beach Island. The island had been created when the 2007 inlet had broken through North Beach, creating what is called a relic island because it is cut off from its upstream source of sand.

Ted, the director of Chatham's Coastal Resources Department and Mark, the Cape Cod National Seashore's Geographic Information Systems expert, didn't like what they saw. The ocean was eroding the front of North Beach Island at an average rate of eighty feet per year.

The "Audubon camp," a small utility shed where scientists from the Massachusetts Audubon Society used to store their bird

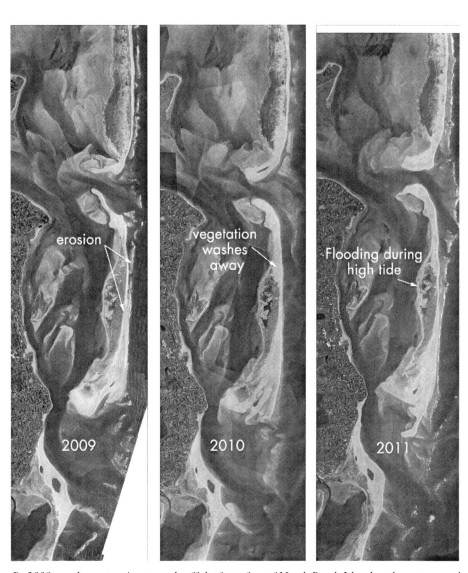

By 2009, sand was starting to erode off the front face of North Beach Island and wrap around on both the north and south ends of the island. By 2010 a large segment of vegetation on the northeast section of North Beach Island had washed away, exposing several camps owned by the Cape Cod National Seashore. By 2011, high tides flooded this area of vegetation. The Cape Cod National Seashore removed its five camps in March 2012. Six privately owned camps still remain on North Beach Island. The island continues to be threatened by sea level rise and could be broken in two by a future storm which would further threaten the mainland. Photo, courtesy of Ted Keon, Chatham Department of Coastal Resources.

banding equipment, was close to being swept away. At this rate, the shed and at least two of the five other camps the Seashore owned would probably wash away this winter, and the rest the following year. Most distressing was the water sweeping into the center of the island from the bayside, threatening the six privately owned camps. Another year's worth of erosion or a major storm could burst the island in two, which would threaten the mainland as well.

Mark reported his findings to George Price, the superintendent of Cape Cod National Seashore. In turn, Price decided to remove all five threatened Seashore camps at once. Doing so would save the government the expense of barging heavy equipment out to the island every time another camp was undermined. He wrote a letter to the Seashore leaseholders telling them they had to vacate their camps by September fifteenth so the Seashore could burn them down and cart away the remains before the end of the year.

Then all hell broke lose.

Legalities

Initially, the leaseholders threatened to publicly humiliate George Price and Ted Keon. This was probably not a wise tactic since their futures lay in the hands of these two competent men. Eventually, cooler heads prevailed, and the leaseholders developed a three-pronged strategy.

First, they tried getting the camps listed as historic buildings. The town of Chatham's wooden bridge was recently accepted for listing on the national register of historic places, as were dune shacks in Provincetown.

The problem was that the North Beach Island camps had all been rebuilt and put on stilts after the 1991 "No Name Storm." Even the camp owners admitted they no longer looked very historic. Not to mention that the Provincetown camps were situated on part of the beach that was growing, while the Chatham camps were on a rapidly eroding relic island. It was doubtful that the keeper of the registry would declare a disappearing island a historic area, but it was worth a try. At least the tactic would give the leaseholders a little extra time as the designation wended it's way through the state and federal bureaucracies.

Next, Bill Hammatt had a local surveyor stake out what remained of his lot on North Beach. His camp and most of his lot had washed away in 2009, but he had erected a flagpole on what he deemed to be a dogleg of his property and stated that as long as the flag was flying he owned the land.

His idea was to declare himself the last man standing on North Beach. He figured that once North Beach started growing again, he and any descendents of his stood to win as much as six miles of oceanfront property within the Cape Cod National Seashore. It seemed to be a gamble well worth taking.

As a lawyer Bill was aware of the case of Annette Appegaard, a North Carolina woman whose father had bought twenty five acres of land on the tip of Topsail Island. By the time he gave it to her the beach had grown to one hundred twenty five acres, with a mile of oceanfront beach worth over a hundred thirty million dollars! Not a bad deal for the daughter of a fourth grade-educated developer who once had rowed his dory across to the island to make a bid on the rapidly-growing beach.

Hammatt's thought was to invite all the North Beach Island camp owners to his property on North Beach once it grew south again. The problem with his grand plan was that Mother Nature would have to cooperate. Right now she was busily chewing away at his tiny dogleg of sand.

The North Beach Island leaseholders' strongest argument for keeping their camps should be based on numbers, but so far the Seashore was the only one who had taken any measurements to determine when the camps would wash away.

I decided to test their measurements. Pacing off the distance from each camp to the ocean and calculating the current rate of erosion, it seemed clear that at least one and probably three camps would wash away during the winter; the other two Seashore camps and the six privately-owned camps in the following years. So, an argument could be made that not all the camps need be removed in one fell swoop.

This wasn't the most cost efficient solution, and maybe it wasn't even rational, but perhaps rationality was not the best response when you were talking about destroying someone's beloved summer home. The republic was not going to fall if the seashore only removed those camps in immanent danger and let nature run her course with the others. It had happened hundreds of times before.

Eventually, the Cape Cod National Seashore Advisory Committee advised Price to postpone his decision until after their next meeting on November fourteenth. Meanwhile, the historic registry application process would buy them until December twenty third.

But Mother Nature was still calling the shots. On August twenty eighth she tore the staircase off the Audubon utility shed during the inland passing of Hurricane Irene. Normally, this would have triggered the town officials to condemn the structure and have the owner remove it before it became a navigational hazard. But a neighboring camp owner had hauled the staircase above the wrackline and out of harm's way. Nobody seemed to have noticed that the camp was starting to break apart, and no official action was taken.

Then, during the high tides and a brief blow on October twenty ninth, the shed was washed down the beach where it lay in a convenient pile of firewood-sized pieces. These magically disappeared in the following weeks while camp owners put in new shelving and lit warm fires in their wood-burning stoves. Nobody said much about defacing government property. They had just saved the Seashore several thousand dollars.

The end of an era. Courtesy of the Cape Cod Times.

Chapter Twenty Six
Winning the Battle but Losing the War
March, 2012

George Price was on the horns of a dilemma. Would the superintendent of the Cape Cod National Seashore go down in history as the man who pulled the trigger too soon by demolishing the five camps it owned on North Beach Island? Or would he be remembered as the man who had waited too long and cost the tax payers thousands of extra dollars by having to clean up the remains of the camps that had been strewn up and down the beach by a winter storm?

Despite criticism from his own advisory board, George pushed forward with plans to demolish the camps all at once. The leaseholders had screamed bloody murder, claiming that the beach would repair itself and the camps would be safe for another twenty years. They wrote to Senators Scott Brown and John Kerry, the head of the EPA, the National Park system and the Interior Department, asking them all to intervene. It was even hinted that the case of the five rental cottages on Cape Cod might land on President Obama's desk.

Some people felt that if the leaseholders had hired an expert witness to argue the simple point that only a few camps were in imminent danger they could have won their stay of execution. Instead, they disputed mainstream science about global warming and coastal geology and pilloried anyone in a position to make decisions about their future.

On March 19, two tugboats towing deep draft barges crept gingerly over the Chatham bar and unloaded powerful front-end loaders, cranes and almost twenty dumpsters onto North Beach Island. The operation was carried out with military precision. Ten armed officers from the Seashore sat in lawn chairs as the front-end loaders smashed into the summer homes. The guards were there because earlier in the week someone had reportedly poured sand in the outboard motor of one of the Seashore's boats and slashed their trailer tires.

But one could not erase the image that this unnecessarily expensive operation looked, for all the world, like recent scenes of military forces bulldozing the homes of innocent bystanders.

It was the end of a long, contentious process. The Cape Cod National Seashore had destroyed its hard-won reputation for working well with local communities by coming across as ponderous, arrogant and uncaring. The leaseholders came across as name-calling hooligans, more intent on vandalism and denouncing decision makers than winning their argument.

It did not have to end this way. But George Price felt constrained by how things had always been done in the parks. The National Park Service had adopted the long-term strategy of essentially letting nature run its course on its lands. In national seashores, that meant removing any structures threatened by rising waters.

At the end of each summer the Seashore would look at the long-term rate of erosion and decide if their camps would be safe for another year. If so, they would issue another annual lease. If not, they would plan to remove them. In the case of the North Beach camps, there was further pressure to remove the camps before April first, the official deadline for when the endangered piping plover returned to nest, and when all this heavy equipment had to be off the outer beach.

The leaseholders had a different perspective. Rob Crowell and Bob Long were living in camps their ancestors had built and used every summer since before the Seashore had come into existence.

When the National Seashore system was established in 1962, it had paid the owners fair market value for their camps and given them twenty-five-year leases to stay in them. These had run out in 1987, and since then, the Seashore had issued leases on an annual basis. It was not a bad deal. The leaseholders paid eight thousand dollars a year to live on public land on one of the most beautiful and expensive pieces of real estate in the country. But the leaseholders still felt the camps were their own private property and knew that if they lost their leases, the loss would reverberate through their families for generations to come.

The other fundamental difference between the two groups was that the leaseholders had more of a short-term perspective. They had lived on the beach every summer and knew it could change from day to day. They didn't believe in arguments about relic beaches and rollover, and figured that the beach had always repaired itself in their lifetimes. Why wouldn't it continue to do so now?

When the inlet that created North Beach Island first opened in 2007, the editor of Chatham's local paper had set up a website devoted to North Beach. He had asked me to use a camera trained on North Beach to make daily erosion reports for the site. This was easy in 2008 and 2009. The inlet was migrating north and the tip of the beach was eroding back at an average rate of ten feet per day. Sitting at my computer in Ipswich, I could actually see individual waves sweep in and tear a foot of sand off the outer beach's coastal dune. Starting as far as two weeks out, I was able to use a simple arithmetic formula to give camp owners an accurate window of vulnerability. In fact I could tell a camp owner exactly how many days he had before his camp would wash away.

The camp owners who were affected earliest used the reports to make sound decisions about when to demolish or remove their homes. This information was important to them, because the Federal Flood Insurance Program would reimburse homeowners if they removed their houses before a storm; if they waited until a storm destroyed their homes, they would have to pay to remove the debris from the beach.

To my regret, the less vulnerable camp owners only learned about the system from the website when their camps were next in line. Rather than viewing the daily reports as the tool they were intended to be, they saw the reports as invading their privacy.

Whether the camp owners liked the reports or not, there was general agreement that you could make short-term predictions about erosion and, in fact, you had to do so in order to plan for your camp's future.

It became commonplace for private camp owners to move their camps only days before they were washed away. Some of the owners moved their camps multiple times, using neighbors' excavators or inexpensive local contractors who became adept at lifting up the camps between two front-end loaders and carrying them up the beach, guided only by two drivers talking to each other on cell phones. It was an impressive operation to watch. One owner had even moved his camp four times within a year before finally barging it to the mainland for safekeeping.

The system had its faults. Eventually, all the camps had to be removed or were washed away. But it did give homeowners a sense of control. They felt like they could act in the face of danger, and if Mother Nature won the battle, at least they had done everything they could against an implacable foe.

It is ironic that if either the town of Chatham, the leaseholders or the Cape Cod National Seashore had mounted a camera on one of the North Beach Island camps all the parties could have at least had daily objective records of erosion or accretion on which to base their arguments, if not to make their decisions.

In retrospect, it is arguable that it would have cost more money to take the camps down individually, but it is also arguable that the Cape Cod National Seashore didn't have to pay an off-Cape contractor close to four hundred thousand dollars to remove the camps all at once. They could have made a series of short-term decisions and used local contractors to remove each camp just before it washed away. Even if one or two of the camps were lost to Mother Nature, it would not have been the end of the world. Such things had been happening for hundreds of years.

Doing so might have preserved the goodwill between the citizens of Cape Cod and the National Seashore, arguably a priceless commodity worth almost any amount of money to maintain.

If there is a single lesson to be learned from the past ten thousand years of history on this barrier beach, it is that while sea level rise is a long-term process, decisions about the effects of that rise can be made by using short-term predictions about erosion. This will become crucial as the three million seven hundred thousand US citizens who live less than a meter above sea level and the two hundred communities protected by barrier beaches have to make similar decisions in the next twenty years of more rapidly rising seas.

In the end, it probably made very little difference in Chatham whether the camps had been removed in 2012, 2013, or 2015. Everyone knew that the era of beach camps was coming to an end.

In the 1800's there were as many as one hundred twenty five people living in simple camps on Monomoy Island, and as recently as the Sixties there had been forty camps on North Beach proper. With the last five private homes removed from North Beach Island, Chatham would no longer have camps on the outer beach.

But that will not mean that fewer people will be able to enjoy the beach. Quite the contrary. For the last few decades more and more people have been taking their own boats or water taxis out to the outer beach. And if you wanted the more intense experience of living on the beach, you could still drive down in a self-contained vehicle to spend the weekend fishing. Naturalists now lead seal and bird watching cruises to the outer beach, made all the more popular by the added thrill that you might just spot a great white shark or whale swimming in this natural environment.

And so, while one era in the ten-thousand-year history of this beach has come to an end, another is just beginning.

Notes

Cover and Acknowledgements

Thomas Hart Benton; A Portrait. Burroughs, Polly. Garden City, NY, 1981. Doubleday and Co.

This book describes the importance of Martha's Vineyard to Benton's style of American realism, and chronicles his first-hand experience of the 1938 hurricane. That experience led him to paint *The Flight of the Thielens.*

In the acknowledgements I used the exact date of the 1938 hurricane. In other chapter headings I have used exact dates when they could be determined and the most probable dates when they could not be determined.

Chapter One, The Beach, June 13, 2011

1. A Natural History of Boston's North Shore. Lindborg, Kristina. Hanover NH, 2007: University Press of New England.

2. Cape Cod and the Islands. Oldale, Robert, Yarmouth MA, 1980: N. Parnassus Imprints.

Chapter Two, The Matriarch, 10,000 B.C.

1. The House on Ipswich Marsh. Sargent, William, Hanover, NH, 2005: University Press of New England.

2. A Geologist's View of Cape Cod. Strahler, A.N., New York, 1966: Doubleday, reprinted by Parnassus Imprints Yarmouth, MA, 1988.

Chapter Three, The Match, November 22, 1600

1. Mayflower: A Story of Courage, Community and War. Philbrick, Nathaniel, New York. 2006: Viking Penguin Books.

2. <u>This Endless Shore</u>. Schneider, Paul, NY 2000: Henry Holt.

Chapter Four, Poutrincourt's Oysters and Squanto's Revenge

1. <u>Mayflower: A Story of Courage Community and War</u>. Philbrick, Nathaniel, New York. 2006: Viking Penguin Books.

2. <u>This Endless Shore</u>. Schneider, Paul, NY 2000: Henry Holt.

Chapter Five, The Sparrowhawk, 1626

1. <u>Sparrow Hawk, Ye Ancient Wreck, 1626</u>. Livermore, Charles; Crosby, Leander, Boston 1865: Alfred Mudge and Sons.

Chapter Six, Never Trust a Bipolar Pirate, April 1717

My interpretation of Samuel Bellamy as suffering from Seasonally Affective Bipolar symptoms is my own conjecture, based on readings in these two excellent accounts:

1. "The 1715 Hurricane Fleet Disaster, Discovery and Salvage." Debry, John, *Historical Research and Development, Inc. Newsletter*

2. "Pirates of the Whydah." Webster, Donovan, *National Geographic* magazine, May, 1999.

Chapter Seven, Haying the Outer Beach, August 10, 1720

Although most of the Chatham salt marsh haying was done on Morris Island and most of the Orleans haying was done on Nauset Beach, I have chosen to situate this chapter on the outer beach to better fit the narrative arc. Nobody is quite sure which exact instance led to the Eldredge/Eldridge distinction, so I decided to place it on the salt hay marsh as well.

Further reading includes:

1. "The 1759-1835 History of Barnstable County." Freeman,

John. This account covers haying and squabbles over the ownership of the Monomoit Great Beach.

Chapter Eight, The "Old" Inlet, 1846-1871

1. "Chatham Lighthouses." Lighthouse, lighthouse.cc/chatham/history.html.

2. Sea Level Rising; The Chatham Story. Sargent, William. Atglen, PA, 2009 Schiffer Publishing.

Chapter Nine, Unraveling an East Coast Secret, The Monomoy Branting Club, April 19, 1863

1. "Fluctuation in Numbers of the Eastern Brant Goose." Phillips, John C., Vol XLIX, 1932: *The Auk,* University of California Press..

2. With Rod and Gun in New England and the Maritime Provinces. Samuels, Edward Augustus, January 1897, Samuels and Kimball.

Chapter Ten, "No Book, No Marriage," The Outermost House, Eastham, 1925

1. The Outermost House; A Year of Life on the Great Beach of Cape Cod. Beston, Henry, 1928: Doubleday, Henry Holt.

Chapter Eleven, Rum Runners and Mooncussers

1. "The Last Mooncusser." Interview by Bill Black, 1979: www.capeirish.com.

2. The Mooncussers of Cape Cod. Kittredge, Henry, Boston, MA, 1937: Houghton Mifflin.

Chapter Twelve, The Conversation, Monomoy Island, 1958

1. Sargent, F.W. personal conversation, 1958.

Chapter Thirteen, The Debate, February, 1973

1. Bertrand, Gerry, personal conversation, 2011.

2. "A Bridge to the Climate Future." Buhl, J. B., April 22, 2011: *New York Times Green Blog.*

3. "The Law's Great Strength." Doremus, Holly, April 22, 2011: *New York Times.*

4. The Endangered Species Act, Wikipedia. www.wikipedia.org

Chapter Fourteen, The Nude Beach Battle, Brush Hollow, North Truro, August 25, 1974

1. Leatherman, Stephen, personal communication, 2011.

2. "Nudity: a Defining Seashore Issue." Milton, Susan, August 7, 2011: *Cape Cod Times.*

Chapter Fifteen, Field Research, Truro, Massachusetts, June 20, 1977

1. Godfrey, Paul, personal communication, 2011.

2. Leatherman, Steve, personal communication, 2011.

3. "Field Report for the Province Lands of Cape Cod." 1977: Earthwatch Expeditions.

4. "The Effects of Off-Road Vehicles on Barrier Beach Invertebrates of the Temperate Atlantic Coast." Steinbach, Jacqueline; Ginsberg, Howard, Cape Cod National Seashore.

5. "The Effects of Off-Road Vehicles on the Infauna of Hatch's Harbor." Wheeler, Nancy, 1979: Cape Cod National Seashore.

Chapter Sixteen, "Everything Has Washed Away," Coast Guard Beach, February 5-8, 1978

1. "The Storm of the Century." Milton, Susan, February 3, 2008: *Cape Cod Times*.

2. Shallow Waters; A Year in the Life of Cape Cod's Pleasant Bay. Sargent, William, Boston 1979: Houghton Mifflin, reprinted by Stony Brook Press, Brewster, MA.

3. "Remembering the Blizzard of 1978." Tougias, Michael. Feb 7, 2008: Cape Cod National Seashore lecture.

Chapter Seventeen, Piping Plover, Symbols of the Atlantic's Great Beaches or Vermin That Should be Shot? May 15, 1086

1. Hecker, Scott, personal communication, 2011

2. "The Curious Case of the Piping Plover." Frieswick, Kris, August 14, 2011: *Boston Globe Magazine*.

Chapter Eighteen The "Old" Inlet, 1987

1. Adapted from Storm Surge; A Coastal Village Battles the Rising Atlantic.Sargent, William, Parnassus Imprints, Yarmouth, MA, 1995. Reprinted by University Press of New England, Hanover NH, 2004.

2. Breakthrough, The Story of Chatham's North Beach. Wood, Timothy, J. , Chatham, MA 1988: Hyora Publications.

Chapter Nineteen, "The Perfect Storm," October 30, 1991

1. Adapted from Storm Surge; A Coastal Village Battles the Rising Atlantic. Sargent, William, Yarmouth MA 1995: Parnassus Imprints.

Chapter Twenty, Life's a Beach, North Beach, 2002

1. Steve Batty, personal communication, 2011.

2. <u>Drifting Memories; The Nauset Beach Camps of Cape Cod</u>. Higgins, Frances L. Old Orleans, MA, 2004: Lower Cape Publishing.

Chapter Twenty One, The New Inlet, April 24, 2007

1. Adapted from <u>Sea Level Rising; The Chatham Story</u>. Sargent, William, Atglen PA, 2009: Schiffer Publishing.

Chapter Twenty Two, Hurricane Noel, November 3, 2007

1. Adapted from <u>Sea Level Rising; The Chatham Story</u>. Sargent, William, Atglen PA, 2009: Schiffer Publishing.

Chapter Twenty Three, The Last Summer, Chatham, Massachusetts, 2009

1. "Paradise Lost; Erosion Claims Last Camp on North Beach's First Village." Pollack, Alan, 2011: *Cape Cod Chronicle*.

Chapter Twenty Four, How to Predict Erosion

1. Adapted from "The Tides of Change." Sargent, William, April 2012: *Cape Cod Life Magazine*.

Chapter Twenty Six, Winning the Battle But Losing the War - March, 2012

1. "Final Eviction Notices in Chatham." Milton, Susan, August 16, 2011: *Cape Cod Times*.

2. "Cape Cod National Seashore camps Again Face Demolition." Milton, Susan, November 23, 2011, *Cape Cod Times*.

3. "Long Time Tenants of 5 Cottages on North Beach Have Until December 31." Milton, Susan, *Cape Cod Times.*

4. "No Erosion of Hope." Schworm, Peter, October 17, 2011: *The Boston Globe Magazine.*

5. "No Time for Dramatic Gestures." Editorial, October 18, 2011: *The Boston Globe.*

William Sargent is the director of The Coastlines Project. To learn more, visit coastlinesproject.wordpress.com.

To learn more about William Sargent's books, visit www.strawberryhillpress.com

Cover image, Flight of the Thielens, *courtesy of Henry Schwob.*
Interior photos by William Sargent, unless otherwise noted.

CPSIA information can be obtained at www.ICGtesting.com
Printed in the USA
LVOW132125270513

335591LV00005B/644/P